Contents

Act in such a way that you always treat humanity, whether in your own person or in the person of any other, never simply as a means, but always at the same time as an end.

Kant

Contributors

Anne McLaren

A graduate and postgraduate student at the University of Oxford, Anne McLaren was Director of the Medical Research Council's Mammalian Development Unit, London for eighteen years. Now based in the Wellcome/CRC Institute of Cancer and Developmental Biology (Cambridge), her research has ranged widely over developmental biology, reproductive biology and genetics including molecular genetics. She was a member of the UK Government's Warnock Committee on Human Fertilisation and Embryology, and is now a member of the UK Human Fertilisation and Embryology Authority that regulates IVF and embryo research in the UK. She is also a member of the Nuffield Foundation's Bioethics Committee and the European Group on Ethics that advises the European Commission on social and ethical implications of new technologies.

Colin Tudge

Colin Tudge read zoology at Cambridge in the 1960s, and became a writer. He also has a lifelong interest in moral philosophy. He contributes regularly to *The New Statesman*, and has published a dozen books on topics such as evolution, genetics, agriculture and conservation. He recently collaborated with the creators of Dolly, Professors Ian Wilmut and Keith Campbell, in a book about cloning. Since 1995 he has been a Visiting Research Fellow at the Centre for Philosophy at the London School of Economics.

John B. Gurdon

Professor Sir John Gurdon was the first person to create normal adult vertebrate animals by cloning in 1958. His scientific career has been devoted to understanding mechanisms of development, essentially through nuclear transplantation, messenger RNA micro-injection and morphogen gradient interpretation.

James Byrne

James Byrne is a PhD student in the laboratory of Dr Gurdon. The primary focus of his research is to elucidate the mechanisms of vertebrate nuclear reprogramming. He has published a variety of articles on the ethical and scientific aspects of therapeutic and vertebrate reproductive cloning.

Keith H. S. Campbell

Professor Keith H. S. Campbell is a cell biologist/embryologist with 26 years' scientific experience, 17 of which have been in the field of cell growth and differentiation, including 12 years working with early embryos. In 1991 he joined the Roslin Institute, where he produced mammalian embryos by nuclear transfer, resulting in the birth of Dolly in 1996. In 1997 he left Roslin to become Head of Embryology at PPL Therapeutics and since 1999 he has been Professor of Animal Development at the University of Nottingham.

Claude Sureau

Professor Claude Sureau is an obstetrician, and formerly Head of the Obstetrics Department at the Baudelocque University Hospital. From 1982-1985 he was President of the International Federation of Gynaecology and Obstetrics, and from 1985-1994 was President of its Ethics Committee. He has also been President of the European Association of Gynaecologists and Obstetricians, a member of the French *Académie de Médecine* since 1978 and was its President in 2000. His publications include: *Alice au pays des clones.*

Egbert Schroten

Professor Schroten studied theology and philosophy in Utrecht and Strasbourg. He was a lecturer in the philosophy of religion and ethics from 1969-1987. Since 1987 he has been a Professor in Christian Ethics at Utrecht University (Faculty of Theology). He is Director of the University Centre for Bioethics and Health Law, Rector of the Reformed Institute for Academic Theology and a member of the European Commission Group of Advisers on the Ethical Implications of Biotechnology (GAEIB) and its successor, the European Group on Ethics in Science and New Technologies (EGE) from 1994-2001. He is also Chairman of the Dutch National Committee on Animal Biotechnology.

Axel Kahn

Axel Kahn is a doctor and a biologist, and is the Director of the *Institut Cochin* in Paris. For ten years he has been a member of the French National Advisory Committee on Ethics, and since 2000 he has been President of the High Level Group on Life Sciences (HLGLS) established by the European Commissioner responsible for Research. He has written numerous works, including: *Copie conforme, le clonage en question*, Nil Edition, Paris, 1998, and *Et l'homme dans tout ça? – Plaidoyer pour un humanisme moderne*, Nil Edition, Paris, 2000.

Dietmar Mieth

Dietmar Mieth is a Professor of Theological Ethics at the Catholic Theological faculty, University of Tübingen. Since 1997 he has been the German member of the European Group on Ethics in Science and New Technologies (EGE) of the European Commission. His numerous publications include: *Ethik und Wissenschaft in Europa. Die gesellschaftliche,*

rechtliche und philosophische Debatte (Ethics and science in Europe. The social, legal and philosophical debate), Freiburg 2000.

Maxime Tardu

Maxime Tardu has a doctorate in law and political science from the University of Paris. He spent his career working for the United Nations Office of the High Commissioner for Human Rights, where he held the post of Director of Research and Treaties until his retirement in 1986. In particular, he led the UN study on "Human rights and scientific progress", which led to the adoption of international standards with respect to people suffering from mental illness and people with disabilities and the role of doctors in combating torture.

André Albert

André Albert has a background in both philosophy and law: having worked as a philosophy teacher, he subsequently became a judge. His current post in the central administrative section of the Ministry of Justice in Paris has led him to specialise in the legal aspects of bioethics and data protection. He has contributed to the Council of Europe's standard-setting work in the bioethics field.

Introduction

by Dr Anne McLaren

Those of us who are keen gardeners are frequently making clones – by taking cuttings, layering, dividing corms, or any other method of vegetative propagation. So why is the Council of Europe publishing a book about cloning, and why have I, a laboratory scientist who has worked all her life with mice and who is at best a very amateur gardener, been asked to write an Introduction?

Mitochondrian: energy producing bodies in the cell, outside the nucleus, which contain a small genome.

This book is not about cloning in plants, which has been practised for millennia, or cloning in frogs and other lower animals, which has been done for decades. It is about cloning in mammals. There are two reasons for this. First, mammalian cloning is a relatively new technology, and the approach with which this book is mainly concerned, and which has stimulated the most controversy, is barely five years old. The second reason, which also accounts for much of the controversy, is that we ourselves are mammals.

So how are we defining cloning?

In this context cloning means making genetically identical animals or plants – identical, that is, for the 50 000 or so different genes in each nucleus of our bodies, ignoring for present purposes the thirty-seven genes in each mitochondrion* outside the cell nucleus, important though they are for generating energy.

One way to produce genetically identical animals is by embryo splitting, physically separating the cluster of cells derived from a single fertilised egg into two or more parts before it has implanted in the uterus. Embryo splitting has been carried out experimentally in laboratory and farm animals. It occurs spontaneously in our own species, to produce monozygotic (or one-egg, "identical") twins. Spontaneous monozygotic twinning can occur not only before but also during implantation, right up to the stage where the primitive streak, the first sign of foetal development, forms on the embryonic plate. Usually there is just one primitive streak, so just one baby. Occasionally, there are two or more, so twins or higher multiples occur (one species of armadillo has four, so quads every time). Sadly, there is sometimes no primitive streak and no pregnancy, or just a tumour. The appearance of the primitive streak marks the beginning of individual (literally "undividable") development. In humans, it occurs about two weeks after conception.

"Identical" twins – identical personalities?

You will find little mention of cloning by embryo splitting in this book, but a great deal about the alternative, more controversial approach, cloning by nuclear transfer. But monozygotic twins are often cited, from two diametrically opposed viewpoints. Those who are concerned to point out how important our genes are (and they are!), point out that monozygotic twins are usually remarkably similar to one another in appearance and even in behaviour and other ways (hence "identical"), sometimes so much so that only those who know them very well can tell them apart. But those who are concerned to stress how important our environment is, both before and after birth (and it is!), draw attention to the monozygotic twin pairs who look less similar to one another, and who are often very different personalities. The conclusion drawn is that "genetically identical" does not mean "identical", thus a clone of an animal or person would not be the same as a copy of that animal or person. This point will be made repeatedly in the pages that follow.

What is cloning by nuclear transfer?

Cloning by nuclear transfer never occurs spontaneously in mammals. Experimentally, an unfertilised egg of the appropriate species is required: its nucleus contains only one set of chromosomes, expecting to be topped up to the full number by fertilisation with a sperm. This egg nucleus is removed, and replaced by a nucleus with the full number of chromosomes, and this reconstructed egg is then stimulated in some way to start developing as an embryo. When the transferred nucleus is taken from an early embryo, the procedure works quite well, since the genes have relatively little to unlearn, and can successfully support development from the very beginning once again. But once cells become specialised to generate all the different tissues of the body, the problem becomes very much greater. Somatic cell (body cell) nuclear transfer (sometimes abbreviated to SCNT) cloning rarely yields more than a few percent of liveborn young. Dolly, the first mammal cloned by transfer of an adult nucleus, was the only lamb born from 277 reconstructed sheep eggs. Often the failures die during gestation, or around the time of birth; others develop abnormalities as they grow older.

What are the potential applications of cloning?

Yet in spite of the low success rates, the potential applications of nuclear transfer cloning in farm animals are vastly important, in particular as a route to genetic manipulation. Campbell's chapter elaborates on this theme.

For developmental biologists, the surprise is not that the success rate is low, but that it works at all. During the course of embryonic development, every tissue lineage has been subjected to innumerable molecular signals, turning off whole batteries of genes and turning on others, gradually narrowing down the options until the final specialisation of the adult body cell nucleus is achieved. SCNT cloning implies that such a nucleus, plunged into the midst of unfertilised egg cytoplasm, has the genetic slate wiped clean, and a new programme imposed that will support the whole of a new embryonic development. As the chapter by Gurdon and Byrne stresses, we know next to nothing about how this reprogramming works, at the genetic or molecular level – but the fact that it does work confirms that the process of developmental specialisation involves no loss of genetic material. Genes may be turned on and off; they are not discarded.

The cloning of human beings – some scenarios for the future?

It is when we turn to the possible applications of nuclear transfer cloning to our own species that the Council of Europe's interest quickens, and the confusion and suspicion felt by the public is most evident. Fancy can (and does) run freely, since we are dealing only with hypothesis. There are as yet no facts, no reality. Somatic cell nuclear transfer cloning has been successful in cats, sheep, cattle, goats, pigs and mice: it has not as yet succeeded in dogs and monkeys, and the few attempts in the human have not looked promising.

The ethics of cloning

Nonetheless, the possible consequences if the methodology worked in our own species arouse such strong feelings that speculation is fully justified. Most of the ethical debate centres around the possibility that nuclear transfer cloning could be used to produce a human baby. Cloning for babies is often referred to as "reproductive cloning". There are many different scenarios, some of which are explored in the chapter by Sureau. Transfer of nuclei from early embryos would be the most likely to work, and might be used to increase the success rate of IVF as an infertility treatment. In principle eight new nuclear transfer embryos could be generated from a single *in vitro* fertilised 8-cell embryo. More ethically dubious would be any attempt to clone an existing individual, whether a dying child or a much-loved parent. Some fertility specialists would be prepared to use SCNT cloning as treatment for sterile patients that were unable to produce any gametes. Different aspects of the ethical debate are addressed by Schroten and by Mieth, with a general ethical overview by Tudge, and a discussion of the legal issues by Tardu. The possibility of international legislation is explored by Albert.

Perhaps the greatest confusion has been generated in the field of regenerative medicine. Great clinical promise is being shown today by research on stem cells. These are cells that are capable of renewing themselves, but also of giving rise to more specialised cells. Stem cells play an essential role in our bodies' repair and regeneration mechanisms. If appropriate stem cell lines could be maintained outside the body, they could be used by transplant surgeons for the treatment of a large number of serious and intractable degenerative diseases, including Parkinson's, stroke, multiple sclerosis, ischaemic heart disease, and certain forms of diabetes. Research is being actively pursued on both human and animal stem cells, with the aim of inducing them first to multiply in culture, and then to differentiate into the required cell type: neural, cardiac muscle, and so on. The stem cells may be derived from adults, from umbilical cord blood, from foetuses, or from early embryos. Embryonic stem cells, which proliferate most readily in culture and are pluripotent, have been extensively studied in mice. For human studies, in countries where human embryo research is legal, embryonic stem cells have been derived from embryos no longer required by couples in IVF clinics for their own treatment, and donated by them for research.

One of the problems in using stem cell lines for clinical treatment is that of immune rejection. Immunosuppressive drugs are becoming cheaper and more effective, with fewer side-effects, but they are not ideal. A number of other approaches are being explored. One hypothetical approach would be to use the patient's own somatic cell nuclei to carry out SCNT cloning, to make the early embryos from which stem cells could be derived. Since the stem cells would be genetically identical to the patient, there would be no risk of rejection. This cloning for stem cells would be different, both practically and ethically, from reproductive cloning (cloning for babies), since the embryo would not be transferred to a uterus. Cloning for stem cells is sometimes termed "therapeutic cloning", although the cloning itself would be in no way therapeutic. As an additional source of confusion, the fertility specialists who wish to use reproductive cloning as therapy for their irremediably sterile patients, also describe their approach as "therapeutic cloning". The chapter by Kahn outlines some of the ethical and other problems associated with cloning for stem cells.

Although I have tried to indicate the main thrust of the various chapters in this volume, the reader will discover that most of the topics are touched on by most of the authors. This leads to a certain amount of repetition and even contradiction, but may perhaps contribute to a well-rounded picture of this complex field. It is our aim to make the reader better informed about cloning; if they also become to any degree less confused, that would be a bonus. What this publication and indeed the whole

Ethical eye series wants to do is to encourage non-specialists as well as those directly concerned in the field to consider other points of view and to take part in the debate. Cloning isn't just reserved for scientists, it has significant implications, in many different ways, for each one of us.

Cloning : who has the right to do what and to whom ?

by Colin Tudge

Ethical discussions of cloning and other high-flown technologies don't seem to get to the heart of things; they tend to be too fine-grained. Of course we must look at the details, the physical realities – of course it matters, for example, whether a particular culture of cells does or does not cohere to form a discrete organism and if so, whether that organism is sentient. Yet the broad principles that should underpin all ethics tend to get lost: principles which, I suggest, evolved in the earliest days of humanity and even before, and have been refined over thousands of years not least within the context of religion. Present-day "ethical" discussion is often wonderfully precise, logical and legalistic, yet the morality that emerges from it is simply that of utilitarianism (who immediately benefits, who is incommoded) or indeed of the marketplace (if someone is prepared to pay, and someone else to supply, then who should gainsay them?). Where the broadest principles are invoked, it is often by representatives of religion, which is fine and proper: except that those representatives are typically concerned mainly with nuances of their own theology, which simply embarrasses everybody else. We learn from such discussion that a follower of Islam might not allow such-and-such a thing, while a Protestant Christian might: which is all very interesting sociologically, but not of great help to humanity as a whole. In this chapter, I want to look into the deep but simple waters, the broadest ethical principles, that lie beneath the surface currents – and in fact are common to all religions; and to see how those principles apply to human cloning and associated biotechnologies.

New biotechnologies and the transformation of humanity

The potential of the new biotechnologies to change the way we live and indeed the way we are – our physical and even mental selves – can hardly be overstated, at least if we consider the time-scale sensibly. Scientists and doctors tend to exaggerate

Genetic engineering: general term covering the use of various experimental techniques to produce molecules of DNA containing new genes or novel combinations of genes, usually for insertion into a host cell for cloning.

Cloning by nuclear transfer: the nucleus of a somatic cell from the animal to be cloned is transferred into a fertilised egg cell whose own nucleus has been removed. The resulting cell is then cultured to form an embryo that is implanted in a surrogate mother.

Cultured cell: cultured cells are grown in a nutrient-containing medium in vessels in the laboratory.

Foetal fibroblasts: fibrous connective tissue cells from a human embryo after approximately seven weeks of development.

1.
Nature Magazine: www.nature.com

2.
Roslin Institute website: http://www.roslin.ac.uk

the speed of progress, typically suggesting that they can "solve" a particular problem within three years, or at best a decade: that is, within the time-frame of whoever is providing their grants. In reality, the Universe is complicated and life is the most intricate of all, and serious biotechnologies take decades or even centuries to unfold. Genetic engineering* was first mooted in the early 1970s and still we are only in the foothills. The technology of vaccination has been with us for two hundred years and still poses enormous problems in human and veterinary medicine, both in practice and theory. The science and technology of cloning are still finding their feet: it is surely grotesque to offer human cloning on the back of a few successes in laboratory animals and farm livestock (and a great many failures, often of a kind that in a human context would be horrific). But this technology and others will still be with us in 50 years, or 100, or 500, and – technically at least – it will not seem so ludicrous then.

Indeed, the general lesson of the past few decades is that there is nowhere to run. In the mid-1980s a very distinguished embryologist commented in *Nature* that cloning of mammals "by simple nuclear transfer"* is "biologically impossible".[1] But barely a decade later in 1995 Ian Wilmut and Keith Campbell at the Roslin Institute near Edinburgh[2] produced Megan and Morag, twin ewe lambs cloned by nuclear transfer from cultured* foetal fibroblasts*, and in July 1996 came Dolly, cloned from cultured mammary gland cells from an adult sheep. But then, when I was at university in the 1960s it was taken as an absolute that genes could not be transferred between organisms that were not of the same or closely related species, because the only way to exchange genes was by sex. Yet the first forays into the technology of recombinant DNA – "genetic engineering" – were made just a few years later, in the early 1970s. In principle, genes can now be transferred between any two or more organisms – human into bacterium, mushroom into human, cabbage into baboon. We have all become members of each others' gene pools. More generally, the expression "biologically impossible" must now be seen to be obsolete. Of course, some genuine biological impossibilities might arise in the future, but it would be cavalier to assume a priori that they will do so. The only safe assumption, now, is that any endeavour that does not break what Sir Peter Medawar[1] called "the bedrock laws of physics" must be considered do-able, not necessarily now, or next week, but in the fullness of time: in decades or centuries. There are no biological barriers, apart from those of physics, to prohibit the wildest of fantasies: nothing except our own sense of what is proper. Legislation alone will not suffice. Laws do not work unless people believe in the principles behind them. Morality and aesthetics, our own instincts, are what matters.

Gamete: sperm or egg cells.

Cloning per se offers two kinds of possibility: to clone entire individuals, and to develop various forms of "stem cell" technology. The first of these has been mooted in two contexts: to provide genetic replicas of favoured individuals (livestock, pets, endangered species, lost loved ones, exemplary human beings); and to help childless couples. Cloning to create genetic replicas seems dubious in most contexts but it does have enormous potential for conservation. For humans, interest has focused mainly on helping childless couples, but these discussions might simply peter out. By the time it is reasonable to offer cloning as a serious clinical service for childless couples (in another half century or so) the technology should already be obsolete. By then it may well be possible to turn ordinary somatic ("body") cells into gametes*, and allow couples to mix their genes by *in vitro* fertilisation.

The second broad possibility – stem cell technology – surely has the potential to transform entire branches of medicine. Most organs of the body contain groups of "stem cells", which are a kind of *in situ* production line. That is, they are able to divide to give rise to more cells like themselves – but also to give rise to many or all the different cells of the tissue of which they are a part. The most familiar stem cells in clinical medicine are those of the erythropoietic tissue, which reside in the bone marrow and divide to produce all the different cells – red and the many forms of white – in the blood. But liver and pancreas and so on have their own particular stem cells too. It is becoming possible to culture different "lines" of stem cells in the laboratory, for example to replace the nerve cells that fail in diseases such as Parkinson's. It should become possible to

1.
Sir Peter Medawar (1915-1987), British immunologist. Joint Nobel Prize winner for Physiology or Medicine, 1960.

provide beta cells in the pancreas – the kind that fail in diabetes. In principle, any disease that results from the failure of particular tissues – which is a very long list indeed – might be helped by such technology. At present the technology runs into controversy because it tends to involve cells from embryos, albeit very young embryos that surely are far from sentience. In the fullness of time, though, it will be possible to culture just about any kind of cell, and to reprogramme it *in vitro*. In half a century or less, it may well be possible to repair a damaged liver, say, by reprogramming stem cells cultured from skin. Such procedures seem ethically innocuous: just good, if very clever, medicine. But scientists will not achieve such refinement unless for the present they work on cells derived from embryos in which the problems of reprogramming are innately easier. The general point is, though, that some ethically charged procedures seem to have a time limit. As the technology is refined, the aspects of it that are ethically charged may simply become obsolete. The technologies of stem cells and of cloning are not identical, but the two overlap and feed into each other and so must be discussed in the same context.

But cloning per se and its potential for change become most interesting of all when associated with genetic engineering. Advances in genetic engineering have meant that genes may be transferred directly from any one organism into any other – or indeed the transferred gene might be synthesised in the laboratory: an artificial, tailored construct with no precedent in nature. The recipient of the "foreign" gene is said to be "transformed". In their present, foothill days, genetic engineers seek simply to transfer single genes: so that maize, for example, the world's third most important staple after wheat and rice, has been given genes to confer resistance to insects. But as the centuries pass, the "engineers" will surely be able to build entire organisms to an exact specification, with the same precision that Ferrari builds motor cars. These will be "designer" organisms. Already some dream of the "designer baby". The dreams tend to be vulgar: extra tall individuals for basket-ball; extra points of IQ to produce the smartest lawyer in town; the goal is almost always to produce some paragon who is able to out-reach, out-smart and generally out-compete his or her fellows, and grow rich. "What a piece of work is a man!" said Hamlet. Later he had cause to mourn, "That it should come to this!"

But genetic engineering in animals is difficult. At least, it is easy to get new genes – pieces of DNA – into the cells because animal cells, unlike those of plants, do not have thick walls. Traditionally, though (and although the technology is new it is still reasonable to speak of "traditions") an animal could be transformed only by introducing DNA into a very young embryo, preferably at the 1- or 2-cell stage. Such embryos are not easy to come by (at least it is tedious to produce them) and since adding DNA is very hit and miss, transformation has been a chancy and time-consuming business. This is where cloning technology comes in, and why cloning is even more important than it may seem. For DNA may now be added to cells cultured in a dish – thousands or many millions of them. It is possible to see which of the cells have incorporated the required DNA into their own genomes; and by nuclear transfer, the same technique that produced Dolly, an entire animal can then be produced from the particular cell that has been successfully transformed. By such methods the biotech company PPL, which collaborates with the Roslin Institute, produced Polly in 1997: a sheep produced in the same way as Megan and Morag, but from cells that had been transformed (and Polly produces a human blood-clotting protein in her milk, for treating haemophilia). Thus, with cloning, genetic engineering in animals has come of age.

In practice, genetic engineers have been delayed too because they rarely knew which pieces of DNA were actually worth transferring; that is, which pieces of DNA corresponded to which genes. This information is now being provided by the science of genomics, of which the Human Genome Project is the most spectacular but by no means the only example.[1] Within a few decades scientists will have listed all the genes in many of the most commercially and medically important species and then, as the techniques of transfer also develop, transformation of animals could virtually become routine (as it is rapidly becoming in some commercial crop plants). When it is routine, some will surely ask: "Why not in humans?"; and then we could enter the age of the "designer baby".

1.
See *Ethical eye: the human genome*, Strasbourg: Council of Europe Publishing, 2001.
http://book.coe.int

Proteomics:
also proteinomics, proteonics. The study of the set of proteins expressed by a genome.

One final biotechnology that might be seen as the icing on the cake is growing apace: that of proteomics*. Genes operate by making proteins, many of which are enzymes; and these enzymes run the metabolism of the body. Crudely at least, genes may be seen as the administrators, and proteins as the executives. Proteomics aspires to work out from first chemical principles why and how particular proteins (including enzymes) function the way they do. This will enable biologists of many kinds (including pharmacologists) to design particular proteins for particular tasks; and hence to design the particular strands of DNA that will produce those proteins. Proteomics is already with us, although it will take decades to come to anything like fruition. But again, it will not go away. It is part of our future.

So, cloning is a heavy-duty technology even when considered in isolation; but it truly comes into its own in the company of three more technologies: genetic engineering, genomics, and proteomics. This quartet is offering us or our descendants the power not simply to transform all living organisms but to re-design them from scratch, or any creature we may care to envisage that does not presume to transgress the laws of physics.

The ethical problems even of transforming insentient life-forms such as plants are not trivial. In animals, which think and feel, they are huge. In human beings, they are off the scale of anything humanity has ever had to contemplate. To devise codes of ethics that get anywhere close to what is at stake, we must trade in something more than legalities. We must get down to the most basic principles.

As a way into the issue, we might consider the question that Lenin proposed as the most fundamental in all politics: "Who, whom?". Who, in this crowded world, has a right to do anything at all that affects other people? The new technologies raise this issue in a form that is especially acute, first because they have such enormous potential, and can affect all of us in many different ways, and indeed affect some families to their core; and secondly, because the technologies themselves are understood and can be deployed only by a minority, an élite of experts. To what extent should those of us who are not specialists hand over our affairs to the tiny minority who are? To what extent, and in what circumstances, should those experts offer and ply their expertise? The issue, in the end, is one of mandate.

Rule by expert: a matter of mandate

The vital matter of "mandate" is illustrated in its simplest form not in the context of medicine, but of agriculture – by the genetically modified crops (or GMOs, where "O" stands for "organism") that have caused such furore in the UK in particular. Mandate is a vital issue in medicine too; but medicine also raises many other matters of welfare, happiness and indeed of "soul" which, we assume, do not apply to maize and barley, and so the ethical waters become even more cloudy.

People in the UK objected to GMOs – indeed they sabotaged field trials – for a variety of reasons, some of them clearly muddle-headed, but some decidedly not. The broad point, which applies in all contexts, is that humanity tends to be dragged along by technology: we and our fellow creatures adjust our lives to whatever comes along. A sea change is required in the way we control technology: we should first decide what kind of life we want to lead and what world we want to live in, and then direct technology towards those goals. In the case of the GMOs, many (including me) feel that that there is a great deal wrong with the industrialised agriculture that is now called "conventional", and that it is time to change direction. The GMOs on offer in Britain's fields were not intended to solve the problems of world food supply or of environment or welfare (even though they were presented in that vein) but to reinforce the position of particular agri-businesses, of the kind that have created the present indus-trialised methods. In short, some people, including me, feel that genetic engineering has a great deal to offer to crop improvement, and probably in particular in poor countries: contrary to fashionable belief, the highest technology some-times has most to offer to the poorest people. But context is all. The goal should be defined and approved first, and only then should the technology be deployed.

In practice, discussion over GMOs tended as always to focus on details, rather than on principles: whether the particular crops are safe, nutritionally or environmentally. In short, inevitably, what should have been an ethical debate devolved simply into risk-benefit analysis. Risk-benefit analysis is certainly pertinent, as I will discuss later. But it is not the end of the matter. The more fundamental issue is why the scientific experts who created the GMOs, and the politicians who allowed the trials to go ahead (and clearly supported the technology itself and the industrialised approach behind it) think they have a right to take any risk at all. The experts, scientific and political, simply have no mandate to do this. If the trials went wrong, then people at large would suffer, either nutritionally or through the further loss of some wild creature. If people at large (known peremptorily in such debates as "the public") had specifically asked the biotech companies to develop the new crops, then of the course the companies would have a perfect right to go ahead. If the companies had first told people at large what they wanted to do, and why, and asked permission then – if permission were granted – they could have gone ahead. But they, and the politicians, were not given a mandate, and did not ask for one. They simply assumed that they had the right to take matters into their own hands. At the very least this is bad manners, and if things did go seriously wrong, it should be seen as criminality.

In fact, the biotech companies argue that they have got the necessary mandate. No one objects, after all, if they breed new crops by conventional crossing and selecting. Yet the new, "engineered" crops are no less "artificial" (they claim) than those produced by traditional breeding. Engineering, in short, is just a technical extension of ancient practice. The mandate that covers breeding therefore covers engineering. Surely, though, this is not so. Conventional breeding, give or take a few refinements, is limited by biological barriers: new genes are obtainable only from within species, or closely related species. Genetic engineers in principle need recognise no barriers apart from those of physical law. There is a clear qualitative difference. At least, anyone who cannot see that this is so should not be taken seriously in ethical debate.

Some point out that there are intermediate techniques, like those of chromosome transfer in wheat (which dates from the 1960s) or induced mutation. But qualitative transition does not imply a dearth of intermediates. There is a continuous chain of intermediates between us and the Devonian fish who were our ancestors, but that does not mean there are no interesting qualitative differences between us and them. Genetic engineering is very clearly a new ball game and it demanded a new mandate. The biotech companies did not seek such a mandate, and deserve to be taken to task. The elected politicians who felt they should simply put their weight behind those companies and deride the people who elected them have lost sight of what democracy ought to imply.

The same principle of mandate of course applies to medicine, but with many other connotations. The paradigm is the relationship between the individual patient and his or her physician. The patient should be in charge. The patient should decide that he or she is unwell, and wants to be treated. He or she then approaches the physician. Physicians may warn people at large that they ought to seek advice – preventive medicine is good, of course – but it not their brief to impose their will on patients. The physician does not immediately say to the patient: "You must have your blood pressure checked, and if it is high or low it must be corrected". The physician, traditionally, acts only when invited; and when invited he or she traditionally asks: "What ails thee?", and the patient defines what is the matter. The physician then offers to help if they think that the matter is within their expertise; providing no guarantees, but promising to do their best, while explaining all the known pitfalls and warning that there may be others that are not yet known. Then, and only then, the patient and physician have a contract, to treat what the patient perceives to be wrong. The physician is sworn to do what is within his or her capability while the patient, who issued the invitation, remains ultimately responsible for what might go wrong. The relationship and the underlying principles are ancient, and were indeed encapsulated by Hippocrates. It is, in the end, a matter of mandate.

Phenylketonuria:
congenital absence of phenylalanine hydroxylase (an enzyme). Phenylalanine accumulates in blood and seriously impairs early neuronal development. The defect can be controlled by diet and is not serious if treated in this way.

Cystic fibrosis:
a lung disease that causes the production of thick mucus in the lungs, hampering breathing.

But this is clearly different from the kind of scenarios that are now envisaged in the context of cloned or designer babies, or even of some forms of gene therapy. A baby, or indeed an embryo, obviously cannot enter any kind of contract. The decisions must be made on his or her behalf – and who has the right to do this? The parents or putative parents perhaps, but how far can their intervention go? At least in the western tradition of medicine, patients seek help only when they perceive some ailment within themselves. In western medicine at least, distinction is drawn between the correction of discrete pathology, and improvement on what is already adequate. The former is physic, the latter is tonic. Organisers of athletics events recognise the distinction: insulin to correct diabetes is acceptable (some great Olympic athletes have been diabetic) but steroids to add gratuitous muscle are not. By what precedent is it reasonable to admit the designer baby, the putative basketball player or super brain, into the arena at all? If we take the established doctor-patient relationship as the paradigm, it is a non-starter.

We may concede, on common-sense humanitarian grounds, that a doctor should treat some genetic defect in a baby who is too young to make decisions, if the defect is treatable, and if non-treatment is known to lead to frank pathology. Sometimes the treatment of the "genetic disorder" is non-genetic: phenylketonuria* is treated by dietary means. However, even if direct genetic intervention were involved, (as it might be in the future in cystic fibrosis*), then common-sense humanitarianism would still give the go-ahead. In this case, responsible adults very reasonably assume power of attorney on behalf of the not-yet articulate baby, in what remains a traditional patient-doctor contract. The distinction between this and the designer baby is again too obvious to dwell upon.

There are some limits and caveats to the doctor-patient contract. It is not adequate when third parties are directly and seriously involved. In such cases, those third parties have a right to a say. Traditional societies placed some restraint on the carriers of plague, smallpox, or leprosy, and the principle surely applies today to transmissible infections. Similarly, proposals to treat cystic fibrosis by applying corrective DNA by aerosols raise ethical issues because, in principle, passers-by might also pick up the gene-laden aerosol. The chances are slight, but again that is not the issue. Passers-by should not be placed at risk at all – unless they give their express permission: unless they issue a mandate. Then again, some religious groups (and others) can and do argue that there are some favours that the patient has no right to ask for, even on his or her own behalf. Euthanasia is still a difficult issue, for example. Many frown on suicide or even on any form of self-mutilation, not least because they feel, as D. H. Lawrence said in another context: "We do not belong to ourselves".

In general, though, the traditional doctor-patient contract takes us a very long way: it shows why some biotechnical possibilities are acceptable, and why others are beyond the pale. In general, too, the doctor-patient relationship, like that of GMOs, is one of mandate. No expert, no politician – nobody at all in fact – has a right to take any risk at all on our behalf unless specifically invited. To do so is to flout the principle of governance made explicit by Thomas Hobbes in the seventeenth century. He said that monarchs (but we can apply the principle to all who aspire to influence our lives) have no right to anything unless a specific contract is entered into; unless the monarch has a mandate.

Yet this principle is constantly flouted. Experts and politicians not only take risks without mandate or invitation but also, adding insult to injury, suggest that those who protest are driven only by ignorance, superstition, or a craving for public attention. If only we ("the public") could perceive the possible benefits, those experts say, then we would also see that the risks are trivial. Thus they contrive to reduce the entire argument to one of risk-benefit analysis, and thus miss the point entirely. But even the risk-benefit arguments, as typically presented, are deeply flawed.

Hazard, risk, and the limits of human understanding

It is easy for experts to make light of other people's fears. Why should a particular gene introduced by a genetic engineer spread more easily to wild plants than any of the tens of thousands of genes that the crop contains already? Realistically, what damage might be wrought by such a gene? Most ecosystems worldwide are already shot through with entire organisms introduced as "exotics" from foreign places; from groundsel in Hawaii and gorse in New Zealand to rhododendrons in Welsh woodlands. What difference would a few more genes make? Or, in the medical context, why not clone a baby, or add a few helpful genes here and there – if not to make a basket-ball player, then perhaps to protect against Aids?

Indeed, Aids might best be tackled at the level of the DNA. This should certainly be looked at; indeed it might be seen as a sin of omission not to do so. But the general argument does not work. We cannot simply brush the caveats aside, as if all protest was hysterical. In particular, it is abundantly clear that all forms of technology, even the most tried and tested, do not always produce the results that are expected. Civil engineers bring a great deal more experience and data to bear upon their bridges and their office blocks than genetic "engineers" are able to bring to their novel crops, or indeed to babies. In truth, genetic "engineering" is a bad metaphor. It is much more like genetic gardening: light the blue touch paper and retire (as I argued in my book of 1993, *The Engineer in the garden*). But however precise the civil engineers may be, however established the basic physics (it mostly comes straight from Archimedes and Newton, after all) every now and again (surprisingly often, in fact!) their bridges fall or at least wobble, like the new Thames footbridge by St Paul's, and their buildings collapse (or shed their giant plate-glass windows like snowflakes, as on one famous occasion in Boston, Massachussetts in the 1970s). The reasons are clear: it is theoretically impossible to predict all the exigencies that the bridge or the building might encounter. Some novel combination of factors might arise that the engineer simply didn't think of – and could not have thought of.

There is a deeper point than this, too. High technology, by definition (or at least by the definition that I like to promulgate) is the kind of technology that is rooted in, and depends upon, science. Medieval windmills are not high tech, wonderful though they are, since they were built without theoretical knowledge of aerodynamics. But modern aeroplanes certainly are high tech, because they are built with such knowledge.

The fact that a particular piece of high tech performs as expected to some extent vindicates the scientific theorising that gave rise to it. But not entirely. Thus Wilmut and Campbell produced Dolly on the back of ideas about the possibility of reprogramming the genomes of cells that had differentiated in culture. Dolly was a success – so does this vindicate the underlying theory? Only up to a point – as Professor Campbell is the first to admit. The cloning of Dolly probably worked for the reasons he thought that it worked. On the other hand, as he points out, other scientists in other laboratories did carry out more simple forms of cloning in the 1980s, and to some extent (as can be seen with hindsight) the theory on which they based their techniques was undoubtedly flawed. What they thought was happening, was not; and what in fact was happening, they did not at the time suspect. In short: the success of a particular piece of high tech does not and cannot vindicate the underlying science beyond all reasonable doubt. Therefore, if you try out a totally novel technology on the back of a novel piece of science, it might go off in directions you simply had not dreamed of. That is always a theoretical possibility.

But there are worse difficulties even than this. Scientists at any one time tend to have the illusion that they understand the world – at least in principle: just a few "Is" to dot and "Ts" to cross and then we'll know all that is really worth knowing. Such claims have often been made, and have always proved ludicrous. The point is not to laugh at our over-confident forebears, but to note the general lesson: that at any one time there will always be areas of ignorance, and, much more to the point, that it is theoretically impossible, logically impossible, to gauge the extent of that ignorance. It may be that modern science effectively floodlights the whole universe, give or take a few shadows. Or it may be that science so far has simply illuminated a few meandering paths across the darkness. Looking out from the areas of illumination, it is impossible to see the difference.

Such problems are disturbing enough when applied in the context of GMOs. They become truly horrifying when applied to the designer baby. We might compare each gene to a word-which has a specific meaning, and so on, and thanks to the Human Genome Project, we will have a complete dictionary. But if genes are words then the genome as a whole is language, and language is more than a string of words. It has syntax, wit, puns, cross-references, colloquialisms, redundancies, allusions to the past. A language works as a whole. The language of the human genome is at least as esoteric – surely more so, by orders of magnitude – than, say, medieval Chinese or Linear B. Would you undertake to edit an epic poem in medieval Chinese if all you had was a somewhat cursory dictionary? Of course not. Neither would anyone who was halfway sane. Yet that is what would be implied if we took the notion of the "designer baby" literally. The possibilities for error are obviously prodigious; and the slightest incongruity could produce a monster. Yet as a simple footbridge may tell us, it is theoretically impossible to anticipate all the possible hazards, for we cannot tell what "nature" is really like until we look. With the designer baby we lose both ways. If the technologies perform only as well as all technologies seem bound to do, then there would be many a hideous, and to a large extent unpredictable, disaster along the way. Indeed the genetic manipulation of babies will always be hazardous – as all technologies are: which of course is why Ferrari, for all their experience, brilliance, and precision, must still employ test-drivers. But we cannot ditch failed babies in the way that mechanics scrap failed cars. On the other hand, if the technology succeeded beyond all reasonable expectation and precedent then our descendants could, in principle, design the present-day rough-and-ready but altogether wonderful *Homo sapiens* out of existence. Whether a species is wiped out by some ecological disaster, or evolves into something else (or in this case transforms itself into something else) it disappears – and extinction is extinction.

All in all, then, we know we have a body of genetic theory, and very wonderful it seems and undoubtedly is. On the back of it biotechnologists are already growing GMOs in the field, and marketing them; and others contemplate cloned and designer babies. What astonishing self-confidence! We can all of us envisage a shortlist of possible disasters – and the biotechnologists cannot say a priori that those disasters will not happen. Much worse, we should raise the theoretical possibility that a whole swatch of disasters might ensue which, at present, we cannot envisage at all. We cannot know in advance all the science that might turn out to be pertinent or know, as a matter of logic, how much of what we ought to know we simply don't know. But evidence abounds that even when we think we do understand the science, and the technology is well tried, things go wrong. Bridges continue to wobble.

It ought to be obvious, then, as a matter of common sense (backed up by logic) that we must proceed with caution: the "precautionary principle" should apply. Of course, if our ancestors had never taken chances, then we would still be living in caves (assuming we had got that far); and caves are not attractive. This is where we must modify or tighten up the general principle of "caution" and think more precisely of "risk-benefit analysis". It is not a sufficient exercise, but it is necessary.

Risk-benefit analysis applied to GMOs or to the genetic engineering of babies gives the kinds of results a sensible person might predict: that genetic engineering in some contexts does indeed seem to have a lot to offer – the discernible benefits sometimes seem to outweigh the risks, even the unknowable risks. But this emphatically does not suggest that genetic engineering can be deployed lightly, in any instance when the benefits are not absolutely obvious, and the risks relatively slight. Thus it seems clear that farmers in the Sahel region of Africa would benefit enormously from a mildew-resistant sorghum. Sorghum is the staple, and mildew commonly takes half the crop, and fungicide is too expensive and brings problems of its own. But it seems impossible to breed a truly mildew-resistant sorghum by conventional techniques because the sorghum gene pool contains no suitable genes. Genetic engineering is necessary, to introduce a gene from some other grass. Yes, there are risks: but in this case, the technique could save a great many lives, and an entire way of life. The herbicide-resistant rape that was being tried in the UK offered no comparable advantages. It would merely have clipped a few fractions of a penny off the price of a commodity that is already cheap. But the risks would be at least as great as in the Sahel.

Then again: it really does seem worthwhile to apply the techniques of genetic engineering to repair the tissues of children with cystic fibrosis, as has been mooted since the 1980s. Genes introduced *ad hoc* would not be passed to the next generation, although most children with cystic fibrosis could safely reproduce (if they were healthy enough) because, although their offspring would be carriers, grandchildren that did not carry the mutant gene at all could be selected at the embryo stage. If third parties who might be infected by the use of corrective DNA aerosol give their permission, then this seems a benevolent and low-risk use of a truly wondrous set of technologies. But to introduce genes that could be passed to the next generation and to all generations beyond; and to do this, furthermore, when the person has no specific, damaging pathologies, simply in the hope of adding a few IQ points to the dynasty, seems a serious risk.

Yet, I suggest, the principles that could guide the new technologies are straightforward. They are ancient, and widely – almost universally – acknowledged.

New high tech and ancient morality

When Dolly first became known to the world in 1997 people at large began naturally enough to speculate on human cloning. Several medical doctors and at least one physicist are already planning to offer a clinical service. Many biologists have entered the ethical fray and some at least – though not Wilmut and Campbell, the principal players – are clearly sanguine about it. One well-known professor said he would like to be cloned out of curiosity, as if this were justification enough, and another said that human cloning "raises no new questions of ethics". He challenged the world to show otherwise. Let's see if we can rise to the challenge.

To begin with, cloning a human being or conferring novel genes upon the next generation certainly raises the ethical ante. The reasoning is simple. No one can be held morally responsible for eventualities over which he or she has no control. (This is not universally accepted – for example, not by those who believe in the doctrine of original sin – but it is a good common-sense rule of thumb that is certainly recognised as the basis of law.) On the other hand, what you can control, you should take due care over.

Normally, people have only very limited control over the genetic makeup of their own children. We all of us exercise mate choice – and it seems morally proper to do so. At least, most of us would consider that it was irresponsible to produce children in partnership with a dangerous psychopath, if the psychopathy was thought to be genetically rooted. In detail, however, the genetic makeup of our own children is outside our control. It depends on the vagaries of meiosis* and genetic recombination*, and which gamete meets which; and if anything is in the lap of the gods then it surely is this. So, we are responsible for the genomes of our children insofar as we can, should, and generally do exercise mate choice. But after that, if anything goes wrong, we really cannot be held to be morally accountable (even though, distressingly, people often do feel guilty when their children, out of the blue, suffer some genetic setback).

But if we clone a baby, or if we engineer the embryo *in vitro*, then we are prescribing its genes. What we presume to prescribe we are responsible for. No one believes in "genetic determinism" but it is true nonetheless that everything that happens to us, the bad as well as the good, to some extent is rooted in our genes. To prescribe another person's genes is to some limited but significant degree to prescribe their lives. It makes us morally responsible for that person's fate and welfare to an extent that is quite outside the experience of all previous humanity. Again, I am inclined to suggest, if that is not a new moral scenario, then it is hard to know what would be.

Clearly (although this is an aside) there are many psychological considerations. Even if the designer child were eminently

Meiosis:
the process by which germ cells (i.e. those in the ovaries or testes) divide to produce gametes. In meiosis I, homologous chromosomes exchange genetic material. In meiosis II the two resulting diploid cells (i.e. which contain two sets of chromosomes) with their recombined chromosomes divide further to form two haploid gametes (i.e. which contain only one set of chromosomes).

Genetic recombination:
this occurs during meiosis and generates further variation between gametes. Homologous chromosomes exchange parts and thus produce new combinations of genetic material.

successful – a star at basket-ball, the brightest lawyer ever – he or she would still have grounds to feel fiercely resentful. Many children are angry with their parents simply for sending them to what they perceive to be the wrong school, however well they fare subsequently. How much more aggrieved would they be if their parents had prescribed their genes! They could well feel that they had been robbed of their individuality. Intelligent children born by artificial insemination (AI) have been known to say that they do not feel quite "real". Designer children might well say this with interest – and however "irrational" the rationalists might tell them this is, the feeling surely would not go away. People who suffer such feelings may be counselled, but they cannot simply be talked out of them. The parents of the genetically enhanced lawyer might well be shocked to find that their super-bright child sues them for all the mental anguish (s)he has been put through. A few such cases would surely take the edge off the technophilia.

To return to the ethical challenge, we might simply point out that it is at least premature to suggest there will be "no new ethical principles". It is impossible to predict what new scenarios will arise – including the strange turns of psychology, even in successfully cloned or tailored children – just as it is impossible in principle to predict precisely the outcome of new high technologies. This is unknown territory, and we will just have to wait and see. There are no a priori statements to be made.

Or – and this I find most interesting – we might simply point out that cloning will raise no more principles because there are no more deep principles to be raised. The deepest principles of ethics which most of us acknowledge are probably at least in part evolved: adaptations to help us get along with our fellow human beings. Those deep principles have been made explicit, time and time again, not by professional moral philosophers (though many have certainly been helpful) but by prophets, the various representatives of the great religions: Moses, Jesus, Mohammed, and such Hindu luminaries as the nineteenth century mystic Ramakrishna. Their approach was not to practise formal philosophy, but to seek what they took to be truth by revelation. Their method has invariably been to seek solitude, and contemplate in tranquillity.

Out of such contemplation three great principles have emerged, which have most succinctly been summarised by Ramakrishna. The first is that it is good to be personally humble (a virtue also emphasised by Aristotle). The second is that we should have "respect" for fellow human beings and for fellow, sentient creatures. The third is that our attitude to the Universe as a whole should be one of reverence.

This is not the place to discuss whether these notions do in fact represent "revealed truth": whether, as the prophets themselves believed and maintained, they are the literal word of God. Pure pragmatism is enough to suggest that as general statements of attitudes of mind, they work. All ethics in the end is rooted in feelings – emotional responses – as David Hume pointed out in the eighteenth century.[1] The arguments of moral philosophers are secondary: to tease out and explore the motives behind the emotional responses, and the likely consequences of actions that are based on them. Feelings drive morality, while the intellect merely talks about it. It is religion, rather than formal moral philosophy, that seeks directly to refine and cultivate the emotions on which the ethical arguments are based. In secular societies the underlying attitudes are left to chance – even though they underpin the entire ethical fabric. Secular societies have thrown out formal religions, and so they have thrown out the prophets who represent and largely define them. But those prophets, collectively and individually, have provided what seem to me to be unimprovable principles. The ultimate source of their wisdom I am happy to leave to the theologians. But the content – humility, respect, reverence – seems to me to say almost all of what needs saying, in all contexts.

Thus, if we were personally humble, would we think of cloning ourselves? If we had such humility, and truly had respect for others, would we for a second entertain the idea that we might impose our own taste in genes upon our children? For although we call them "our" children for convenience, there is no ownership. They are their own people. If we truly regarded the Universe as a whole with reverence, and the life it contains, would we take such risks with our fellow creatures, just to knock a few pennies off a tonne of maize? Surely not. Surely if

1.
Hume, *A Treatise of nature* (1739), reprinted 1978, (eds. L. Selby-Bigge and P.H. Nidditch), 2nd edition, Oxford: Clarendon Press.

we just remembered the simple roots of morality, as spelled out over the past three-and-a-half thousand years – and undoubtedly were acknowledged for many thousands of years before that – we would deploy the new technologies with a much surer touch. It would still be necessary to frame careful laws and codes of practice (for example on the culture of embryo cells for tissue repair) but the general shape of those laws, what they and we should be trying to achieve, would be obvious.

I feel that people at large know all this perfectly well, even if the arguments are not always made explicit. It's the scientists and philosophers steeped in their own specialities, or at least many of them, who seem to have trouble. This is yet another reason for not accepting rule by expert; and why experts would do well to listen, far more than they are inclined to do, to what their critics are actually saying, and not assume that it is all the baying of ignorant people. The notion that all criticism springs from lack of appreciation may be comforting, up to a point, but it just isn't so.

The history of cloning

by Professor John B. Gurdon and James A. Byrne

"Cloning" is an ambiguous term, even in biology. It basically implies the production of genetic copies. However, these "genetic copies" can be strands of DNA*, cells in cell culture* or entire organisms, and the procedure is still referred to as "cloning". Even the latter process (the cloning of organisms) can be describing bacterial proliferation, the cloning of plants from cuttings (as has been practised for thousands of years), or the cloning of animals. In this chapter we review the history of animal cloning.

The chapter has been divided into three sections. In the first part of the chapter the early history of amphibian cloning is reviewed, as scientists endeavoured to answer fundamental biological questions surrounding development and differentiation*. Part two explains how cloning research expanded into mammals, and the various landmark achievements that have been achieved in this field. In the last part the possible implementation of this technology in humans for both therapeutic and reproductive purposes is discussed.

DNA:
deoxyribonucleic acid, the universal hereditary material.

Cell culture:
the growth of cells *in vitro*.

Differentiation:
the progressive restriction in possible cell fates, until only one cell fate is left.

Somatic cells:
all body cells that are not part of the germ line.

Zygote:
the cell resulting from the fusion of an egg and a sperm (i.e. a fertilised egg).

Amphibian cloning

Several concepts and definitions should be explained before we delve into the history of vertebrate cloning. Vertebrates are animals that possess a backbone, which includes amphibia and mammals. Vertebrate cells are composed of a nucleus where the genetic information (DNA) is stored, and a surrounding fluid called cytoplasm. The cells that produce sperm and eggs are called "germ line" cells, those that comprise the rest of the tissues of an animal's body are called "somatic"* cells. When sperm and egg cells combine (fertilisation) the resulting fused cell is called a "zygote"*. This zygotic cell divides (cleaves) into a ball of "daughter" cells called an "embryo". This embryo can then develop over time into a fully formed organism.

Ever since scientists tried to apply themselves experimentally to the problem of how a fertilised egg turns into a fully formed

Gene:
a stretch of DNA on a chromosome that codes for a protein (and thus possibly a characteristic, e.g. brown eyes).

Germ line:
any cell in the series of cells that eventually produce sperm and eggs.

Blastula:
an early embryonic stage in animals, consisting of a hollow sphere of cells..

Cell membrane:
membrane that surrounds the cell cytoplasm and nucleus.

organism, it has been clear that the primary question concerns the role of the nucleus and its genes*. In the late 1800s, Weismann speculated that genes might be lost in all cells other than the germ line* (Weismann 1892). He suggested that as the early embryo generates the different kinds of somatic tissue (such as skin, intestine and muscle) then genes no longer needed would be discarded. For example, the lineage of cells that gives rise to muscle tissue would lose the genes needed for making skin or intestine. Likewise the cells which form intestine would lose the genes needed for making all tissue types other than intestine.

The first serious attempts to test this idea go back to Spemann who in 1914 succeeded in tying off a small part of the cytoplasm of a newt egg using a thin thread. When the main part of the embryo had reached the 16-cell stage (that is, a ball of 16 cells) he loosened the loop and allowed a nucleus from one of the cells to migrate into the bulge of cytoplasm. He then tightened the loop so that the piece of cytoplasm (with the migrated nucleus) was separated from the rest of the embryo. Spemann found that this piece on its own could form a normal embryo. Therefore, Spemann was able to conclude that (at least up to the 16-cell stage) there was no loss of genetic material (genes) from nuclei. However, this still left the possibility that genes were lost after the 16-cell stage.

It was already clear to Spemann and others that the ideal experiment would be one in which the nucleus of a somatic cell could be transferred into an egg whose own genetic material had been removed (Spemann 1938). This is, in principle, the basis of a nuclear transplantation experiment.

The first real success in achieving nuclear transplantation in animals was that of Briggs and King in 1952, working with the American frog *Rana pipiens*. They made a very fine pipette such that the cell of a blastula* (early embryo) could be sucked gently into it, breaking the cell membrane* (that surrounds the cell) but leaving the nucleus still covered by its own cytoplasm. They then injected this ruptured cell (with its protected nucleus) into the cytoplasm of the very large enucleated egg. "Enucleated" means that the genetic material (genes) of that

egg had been removed. Briggs and King found that a significant number of the enucleated eggs (that received a blastula
nucleus) could develop into embryos and even young tadpoles
(Briggs and King 1952). This was the first demonstration that
showed it is possible to transplant a living nucleus from a
somatic cell into an enucleated egg and obtain development.

Briggs and King continued their experiments with *Rana pipiens*, using donor somatic cells from more advanced embryos.
They obtained the result, surprisingly, that as nuclei were
taken from somewhat more advanced cells, (that is, those of
gastrula* and neurula* embryos, which are late embryo stages),
the rate of survival of the nuclear transplant embryos
decreased. They found that when nuclei were taken from
neurula embryos (about two days old), the recipient eggs no
longer developed normally. Their results led them to the conclusion that as development and cell differentiation proceed,
there is either a loss or irreversible inactivation of genes. This
provided support for Weismann's "genetic loss" theory.

In their later work, Briggs and King asked whether the developmental defects in embryos reflected the kind of cells from
which nuclei were taken. For example, they asked whether the
nuclear transplant embryos derived from endoderm* (future
gut) nuclei were particularly likely to lack non-endoderm
structures such as muscles and nerve. Although they did see a
slight tendency in this direction, the results were never
persuasive, and it is now generally considered that the types
of developmental abnormalities (from which nuclear transplant embryos suffer) does not relate to the cell type of the
nuclear donor.

Soon after Briggs and King's early work with *Rana pipiens*,
nuclear transplantation work was initiated in the UK by
Michael Fischberg's group. They decided to use the South
African frog *Xenopus laevis* for their work for two reasons. One
was that this species of frog can be readily reared to sexual
maturity in the laboratory. The second reason was that, unlike
frogs of the *Rana* species, *Xenopus* frogs can be made to lay
eggs throughout the year by injecting them with certain

Gastrula embryo:
mid-stage embryo
(comprised of three
layers).

Neurula embryo:
late stage embryo
(neural system
beginning to form).

Endoderm:
part of the early
embryo that forms
the intestine and
associated organs.

Genetic marker:
a distinctive genetic change that allows clear identification (e.g. albinos).

Totipotent:
cell that can become any tissue of the final organism.

Reprogrammed:
the change of a nucleus from a specialised (somatic) state to a non-specialised (embryonic) state.

Wild-type:
the most frequently observed state (e.g. the wild-type frog colour is green).

mammalian hormones. *Xenopus* turned out to provide very favourable material for nuclear transplantation. Furthermore, it was possible to use a genetic marker* to prove, beyond any doubt, that the nuclear transplant embryos were derived from the transplanted nuclei, and not from a failure to remove the nucleus from the recipient eggs (Fischberg, Gurdon and Elsdale 1958).

Subsequent work on *Xenopus* led to a conclusion opposite to that of Briggs and King. It became possible to clone from increasingly specialised cells of the endoderm and those of the gut lineage, and yet still obtain normal development. Finally it was possible to transplant nuclei from the intestinal epithelium (very specialised flattened cells) of feeding tadpoles. Some of these cloned embryos turned into adult frogs and became sexually mature males and females (Gurdon 1962). Therefore, the nucleus of an intestine cell is "totipotent"* (following nuclear transfer) in the sense that it contains all the genes that are required to differentiate into any tissue. This established the general principle that cell differentiation can take place without any irreversible inactivation or loss of genetic material (genes). The nucleus of an intestine cell will have inactive genes for muscle, brain, blood, etc., and yet these inactive genes can be reactivated (reprogrammed*) when nuclei are exposed to egg cytoplasm. Thus, the full range of cell types for all body tissues can be generated from a nucleus which has already differentiated; in this case, an intestinal epithelial cell nucleus. This research proved Weismann's "genetic loss" theory to be incorrect. There is no loss of genetic material (genes) as cells differentiate and specialise.

In later work the same general conclusions were reached using nuclei of other kinds of specialised cell. These included the nuclei of neural cells (Simnet 1964) and the nuclei of adult skin cells (Gurdon and Laskey 1970, Gurdon, Laskey and Reeves 1975). In the latter experiments, albino frogs were used to provide the donor material. Thus nuclei from cells of albino tadpoles were transplanted into wild-type* (not albino) eggs. All of the resulting cloned tadpole embryos were albino, proving that they had developed from the donor albino nucleus.

It has been made evident, from the above discussion, that the original nuclear transplant experiments specifically addressed the problem of genetic totipotency in relation to cell differentiation. This was the motivating force behind such experiments and the work provided an answer to a fundamental problem in developmental biology. It was clear, in these early days, that the method of nuclear transplantation would generate (as a by-product) cloned individuals. By a "clone" we mean a group of genetically identical individuals. We are accustomed to the concept of identical twins in humans. These derive from the separation of an early embryo into two parts, so that each forms an individual with an identical genetic constitution. The nuclear transplantation procedure can generate clones containing large numbers of genetically identical individuals. Thus, if nuclei are taken from a single embryo, larval* or indeed adult animal, and transplanted into recipient eggs, then all of the resulting animals will be genetically identical, and therefore constitute a clone.

Larvae:
a juvenile developmental stage (in frogs this would be the tadpole stage).

In 1961 the first picture of a clone was published (Gurdon 1961). It is relatively straightforward to provide large numbers of genetically identical individuals constituting a clone using embryo cells as donors. Frogs, and in fact all vertebrates, will undergo immune rejection of any transplanted tissue that is not genetically identical to them. Thus, proof of the genetic identity of members of a clone was obtained by grafting skin from one animal to another. In the case of cloned frogs, skin grafts were entirely successful (that is, not rejected).

At an early stage in amphibian nuclear transplantation work, it was found useful to carry out the procedure called "serial nuclear transplantation". This involves transplanting a nucleus from an embryo into an enucleated egg and growing an early stage nuclear transplant embryo. The cells of that early embryo are then themselves used as donors to make further nuclear transplants into another set of recipient eggs. And this procedure can be continued indefinitely. Therefore it is possible, in principle, to create enormous numbers of genetically identical individuals through this serial nuclear transplantation or "serial cloning" technique (for review see Gurdon 1986). Serial nuclear transfer appears to increase the cloning efficiency, partly because of a double exposure to egg cytoplasm (which allows further reprogramming), and partly because the donor cell during the second nuclear transfer is an embryonic cell, not a differentiated (and thus harder to clone) somatic cell.

In conclusion, cloning of vertebrates (as was first done in amphibia) was a by-product of the important scientific question "is there a loss of genes in somatic body cells as development proceeds?" Amphibian research has proved that there is not. Cloning mammals is primarily a subject of practical interest rather than a procedure aimed at answering a scientific question. The recent success of nuclear transplantation in mammals (Wilmut *et al.* 1997) has elicited widespread public interest in this procedure, primarily because it is likely to be applicable to humans. The next section reviews the history of mammalian cloning, and the third section discusses the possible implementation of this technology in humans for both therapeutic and reproductive purposes.

Mammalian cloning

In several important aspects, mammalian nuclear transfer is technically more difficult than amphibian nuclear transfer. Mammalian eggs are less than one-thousandth of the volume of amphibian eggs, produced in substantially lower numbers, and invasive procedures must be employed in order to obtain them and to implant the resulting cloned embryos following nuclear transfer. These reasons help explain why, historically, research into mammalian cloning was initiated at a much later date than amphibian cloning research.

Mammalian cloning actually comprises two fundamentally distinct fields: somatic cell nuclear transfer and embryo splitting. Somatic cell nuclear transfer is what most people refer to when mammalian cloning is discussed, that is, the transfer of a somatic nucleus from one cell into another cell (typically an egg) which has had all its nuclear material removed. The resulting embryo has the potential to develop into a cloned organism, genetically identical to the donor cell. The other form of cloning is called embryo splitting, and is literally the natural or artificial separation of an embryo into two or more partial-embryos. These partial-embryos can potentially survive this separation and develop into genetically identical individuals. This is what happens when monozygotic* twins are naturally created. Although there are two distinct types of mammalian cloning, the focus of this review will be on somatic cell nuclear transfer, a subject that offers a huge array of potential future applications across many fields.

The first attempt to clone mammals by somatic cell nuclear transfer was made by Derek Bromhall in 1975. Bromhall fused early embryonic cells with unfertilised rabbit eggs using Sendai virus. Exposure to the Sendai virus causes fusion of both the nuclear donor and enucleated recipient cells. The cloned embryos all died at early embryonic stages (Bromhall 1975). Thus, this first attempt at mammalian cloning had little success. Retrospectively, such problems may have been avoided if another species had been used.

Although the focus of this chapter is on mammalian cloning by nuclear transfer, an account of the work that has been performed on embryo splitting in mammals can be summarised relatively briefly. The first successful attempt to clone mammals by artificial embryo splitting was made by Steen Willadsen in 1979. Willadsen managed to artificially separate the blastomeres* of early sheep embryos, which resulted in five pairs of healthy monozygotic twinned sheep (Willadsen 1979, 1981). Since this experiment several other mammals have been cloned via this method, and Jerry Hall has performed embryo splitting even on human embryos (Hall *et al.* 1993). The cloned human embryos that Hall produced were not subsequently implanted. Recently this procedure of splitting human embryos has been declared ethical by the ASRM (American Society of Reproductive Medicine) as a procedure to help infertile couples conceive children (ASRM 2000). This is a very basic review of the mammalian embryo splitting that has been performed; the rest of this chapter will concentrate on cloning by somatic cell nuclear transfer.

Karl Illmensee and Peter Hoppe were the first to claim to have successfully cloned an adult mammal via nuclear transfer. In a paper to the science journal *Cell*[1] (Illmensee and Hoppe 1981),

Monozygotic: derived from one zygote.

Blastomere: one of the cells into which an egg divides during cleavage.

1.
Cell:
http://www.cell.com/

Inner Cell Mass (ICM):
group of cells in a blastocyst that will form the foetus.

Blastocyst:
the hollow sphere of cells that develops from the morula (the solid mass of cells produced by the first divisions of a fertilised egg) and implants in the uterine wall.

the scientists declared that they had successfully produced three live mice by transfer of embryonic ICM (Inner Cell Mass)* cell nuclei into enucleated mouse zygotes (a zygote is just a fertilised egg). However, no group has been able to replicate this experiment to date (McGrath and Solter 1984, Howlett, Barton and Surani *et al.* 1987, Wakayama *et al.* 2000), generating doubt about the validity of these results (Wilmut, Campbell and Tudge 2000).

The first successful and reproducible mammalian nuclear transfer was performed by James McGrath and Davor Solter (McGrath and Solter 1984). They fused a zygotic nucleus from one zygote (the donor) with another enucleated mouse zygotic cell (the recipient) using the same Sendai virus technique Bromhall had used. This research was not actually somatic cell nuclear transfer (as a zygote is not a somatic body cell) but it was the first confirmed successful mammalian nuclear transfer. This experiment resulted in a high proportion of healthy mice developing from implanted blastocysts* (early embryos), but is the only known protocol for nuclear transfer to enucleated zygotes that results in live cloned offspring. We now believe that the zygotic cytoplasm has very poor reprogramming ability, and thus this experiment probably succeeded because very little or no reprogramming of the donor nucleus was required; it was already in a zygotic state. All contemporary mammalian nuclear transfer research has focused on transferring somatic donor cell nuclei into enucleated eggs, which appear to possess greater reprogramming ability (that is, ability to reprogramme a differentiated (specialised) cell nucleus back into an embryonic state).

After pioneering mammalian cloning by embryo splitting, Steen Willadsen decided to move into cloning via somatic cell nuclear transfer. In contrast with previous cloning research, Willadsen was more interested in efficiently cloning valuable livestock, rather than investigating the biological questions surrounding cloning (McLaren 2000). In 1986 Willadsen performed somatic cell nuclear transfer in sheep, transferring early embryonic cell nuclei into enucleated eggs using a novel electrofusion technique (which involved using electricity to fuse the donor cell and enucleated egg – see p. 179) rather than the Sendai virus technique that had been used previously. This research resulted in several liveborn lambs, and was the first time in history that mammals had been cloned from cell nuclei that were not necessarily part of the germ line (Willadsen 1986).

Although Willadsen's work clearly demonstrated that mammals could be successfully cloned from embryonic cell nuclei, these nuclei were relatively undifferentiated, and most of the potential applications of mammalian cloning required that differentiated or adult donor cell nuclei were used (Gurdon and Colman 1999). This problem was left unresolved until 1996, when Keith Campbell and Ian Wilmut announced that they had successfully cloned two sheep (Megan and Morag) from the nuclei of differentiated cell nuclei (Campbell *et al.* 1996). The donor cells they used were actually failed embryonic stem cell* cultures that had differentiated into epithelial-like (flattened) cells after several passages*. Although embryonic by origin, the passaged cells had a differentiated morphological appearance and expressed markers associated with differentiation. Campbell forced the established cell line into quiescence* (inactivity) by drastically reducing the serum* concentration of the growth media*. This quiescent nuclear state probably increases the efficiency of cloning, but contemporary research has shown that it is not the key to cloning (as was once suggested), as donor nuclei in various other nuclear states have also resulted in viable cloned offspring (Cibelli *et al.* 1998). The actual protocol Wilmut and Campbell followed was very similar to Willadsen's electrofusion experiment, where he proved electrofusion to be more successful than the Sendai virus technique for fusion of donor cell with the recipient enucleated egg.

Following the success of Megan and Morag, Wilmut and Campbell (with a degree of trepidation) decided to try the same technique with adult cell nuclei. They induced a culture of adult mammary cells into quiescence (by reducing the serum concentration) and electrofused these cells with enucleated sheep eggs. Although several hundred eggs were fused with donor cells, only one sheep (Dolly) was conceived (Wilmut *et al.* 1997). Dolly, a fertile adult cloned from an adult cell

Embryonic stem cell (ES-cell): cultured embryonic cells that can proliferate indefinitely and differentiate into many different tissues.

Passage: the process of subculturing a cell line into new flasks and medium (usually because they had outgrown the original flask).

Quiescence: an inactive cell state where division and growth have temporarily ceased.

Serum: the component of a cell culture growth medium that promotes cell division.

Growth media: a solution that cells grow in.

Reproductive human cloning:

production of a human being that is genetically identical to another (by nuclear substitution from a human adult somatic cell or child cell, or by artificial embryo splitting).

Therapeutic human cloning: cloning where the object is not to implant the clone, but where experiments may be carried out on it with particular long-term therapeutic goals, or where it may be used to grow tissues for therapeutic transplantation.

nucleus, was the cumulation of decades of research into vertebrate cloning, both amphibian and mammalian.

Following Dolly, scientists began to clone a variety of other mammals from differentiated and adult tissue. Teruhiko Wakayama produced a clone of fifty mice using a novel microinjection technique to transfer the donor adult cell nuclei into the recipient enucleated eggs (Wakayama *et al.* 1998). Monkeys were also cloned, but only from embryonic tissue (Meng *et al.* 1997), and goats were cloned from foetal tissue in 1999 (Bagushi *et al.* 1999). Also in 1999 Wells cloned cows from adult cell nuclei (Wells *et al.* 1999), and recently three groups succeeded in cloning pigs (Polejaeva *et al.* 2000, Betthauser *et al.* 2000, Onishi *et al.* 2000). But cloning has not yet been successful in many other mammalian species including dogs, rats and horses. These species are mentioned because efforts to clone these animals have been publicly announced. Recently, various scientists have declared that they intend to clone humans via somatic cell nuclear transfer (Zavos 2001), and this has predictably captured substantial media attention.

Human cloning

Having summarised the history of animal cloning with non-human vertebrates, it is perhaps appropriate to discuss the possible future applications of somatic cell nuclear transfer technologies in humans. Human cloning, or somatic cell nuclear transfer with human cells, is actually the foundation for two quite distinct technologies: reproductive human cloning* and therapeutic human cloning*. Reproductive human cloning involves the creation of a cloned embryo using donor cells from the potential parent, which is implanted into the womb, develops into a foetus and is eventually conceived. The aim of this line of research is to reproduce a genetically identical twin to the cell donor. Therapeutic human cloning involves the creation of a cloned embryo using non-diseased donor cells from a patient with a degenerative disease or disorder (for example, Parkinson's). This embryo is not implanted, but is cultured into an embryonic stem cell line (immortalising these cells). These embryonic stem cells can then be chemically differentiated into potentially therapeutically useful cells, such as dopamine-producing precursor cells for Parkinson's sufferers. As the cells would be genetically identical to the patient, we can assume that there would be no immune rejection of this tissue (Gurdon and Colman 1999).

The aim of therapeutic human cloning research is to cure or alleviate the symptoms of various diseases or disorders, which has resulted in significant support from various scientific and medical groups (Kind and Colman 1999). However, because this technology relies on the destruction of a cloned human blastocyst (to create the embryonic stem cell line) various religious and secular groups have voiced fervent opposition (Society for the Protection of Unborn Children 2000). The specific use of therapeutic human cloning to produce cloned embryonic stem cell lines was ratified in the UK by large majorities in the House of Commons in December 2000, and the House of Lords in January 2001 (BBC News 2001).

Advocates for reproductive human cloning enjoy much less support. Aside from ethical (Pence 1998) and religious arguments (Turner 1997), there is considerable scientific opposition to the proposal to reproductively clone humans (Jaenisch and Wilmut 2001). This scientific opposition is primarily based on suspicions that the procedure will suffer from the same problems observed when other mammals are reproduced via somatic cell nuclear transfer.

The first problem observed in mammalian cloning is a consistently low efficiency of reconstituted eggs developing to parturition (birth). Typically, to get one cloned animal to parturition, approximately 100 eggs must be enucleated and reconstituted with donor somatic cell nuclei, either by electrofusion (Wilmut *et al.* 1997) or microinjection (Wakayama *et al.* 1998). Thus nuclear transfer is usually only 1% efficient. Even the highest efficiency observed in reproductive mammalian cloning from adult donor cell nuclei does not exceed 3% (Wakayama *et al.* 1998). If human reproductive cloning suffers from this low efficiency, then large numbers of human eggs would be needed to conceive a single child.

LOS (Large Offspring Syndrome): a collection of congenital defects, most notably oversized newborns..

IVF (*in vitro* fertilisation): biological process that occurs in a container outside of the organism e.g. cells or tissues grown in culture.

ART (Assisted Reproductive Technology): for example *in vitro* fertilisation.

Imprinted genes: copy of genes derived from either the maternal or paternal genome that are suppressed in the embryo.

IGF2R gene : (Insulin-like Growth Factor 2 Receptor gene): this gene codes for a receptor that helps control the size of the foetus.

The second scientific argument against reproductive human cloning is the developmental abnormalities that have been observed in various mammals that have been conceived via somatic cell nuclear transfer. The applicability of these animal data to humans has been debated at a National Academy of Sciences conference in Washington, DC (National Academy of Sciences 2001). Scientists at the conference who oppose reproductive human cloning pointed out that up to a third of the mammals cloned have developmental abnormalities, most commonly a collection of defects referred to as LOS (Large Offspring Syndrome)* and that the same abnormalities would probably occur following human somatic cell nuclear transfer. Scientists also observe that there is currently no molecular technique available that could screen the entire genome for incompletely reprogrammed genes following nuclear transfer, and that any one of the 30 000 + human genes could be incorrectly reprogrammed following nuclear transfer (Jaenisch and Wilmut 2001). Conversely, scientists who support reproductive human cloning suggest that many of the defects observed in animal cloning are due to poor culture conditions, culture conditions that have been improved and optimised for human embryos and cells over the twenty-three years we have performed IVF* and other ARTs* (Zavos 2001). They also note that LOS defects appear to be correlated to incorrect imprinting* of the IGF2R gene* (Young *et al.* 2001), and that that this gene is not imprinted in humans or other primates (Killian *et al.* 2001) suggesting these species may be safer to clone. The difference in incidence of LOS defects following human and non-human IVF provides empirical evidence supporting this hypothesis (Killian *et al.* 2001). However, this does not mean that reproductive human cloning is necessarily safe, as there may be other errors in reprogramming of the other 45 imprinted human genes, or the over 30 000 non-imprinted genes. It has been suggested that inefficient placental vascularisation as well as other problems may still occur if we reproductively clone humans (Wilmut, quoted by Brower 2001). Humans and other primates may be somewhat safer to reproductively clone than artiodactyls (sheep, cows, pigs) and rodents, but "safer" is not necessarily "safe".

In summary, research into vertebrate cloning has deepened our understanding of developmental biology and provided us with a huge array of potential future applications.

As well as the previously mentioned therapeutic human cloning to produce embryonic stem cells genetically identical to a patient, other medical benefits include pharming* and xenotransplantation. Pharming is when a cloned transgenic animal (for example a sheep or cow) secretes therapeutic proteins in their milk (Schnieke *et al.* 1997). Recently this procedure has been extended to chickens, where large quantities of the therapeutic protein can be produced in the egg white. Xenotransplantation is the theoretical procedure where a cloned transgenic pig is used to provide organs temporarily for a transplant patient until a compatible human match can be found (Platt 1999). On the agricultural front, when the cost of reproductive cloning eventually drops, cloning élite livestock will become feasible, allowing farmers to increase the general productivity and health of their animals (Mitalipov and Wolf 2000). On the issue of basic research, genetically identical laboratory animals (for example, laboratory mice) would increase the accuracy of animal research, as the only variability in experiment results would be due to experimental variables, rather than genetic differences. As a tool to aid conservation of rare species threatened with extinction, endangered animals could have their population numbers (although not variability) increased through cloning. In general, animals of special ability, for example high pedigree horses, could be cloned (Westhusin 1998), and in February 2002 a cat was cloned.[1] Finally, cloning could theoretically be utilised for human asexual reproduction, but as previously mentioned, the efficiency and safety of this proposal is in question. It appears reasonable to suggest that the incidence of developmental abnormalities observed in non-human primates (following somatic cell nuclear transfer) would provide an indicator as to the safety of reproductive human cloning. Although initial primate experiments have shown no developmental abnormalities following nuclear transfer with embryonic donor cells (Meng *et al.* 1997), the numbers are too low to be conclusive, and further non-human research with differentiated somatic donor cell nuclei would be prudent.

Pharming: a pharmaceutical version of the more traditional "farming".

1. See: http://www.south-erndigest.com/vnews/display.v/ART/2002/02/22/3c754072134 19

While reproductive human cloning remains theoretically risky and highly controversial, we believe that therapeutic human cloning should be considered separately, and the benefits and disadvantages debated on their own merit and cost. The potential medical benefits of therapeutic human cloning for adults and children who suffer with degenerative diseases and disorders have resulted in many advocating the implementation of this technology as the most ethical position:

> "Through the therapeutic cloning of human embryonic stem cells, we have the potential to cure a huge plethora of diseases and disorders that plague our species. How can a civilised society turn its back on the diseases, sickness and suffering of its own people? Is a microscopic cleaving egg, four days after fertilization, really more important that curing your mother's Alzheimer's, your father's Parkinson's, or your son's diabetes? And before you answer, realise that the diseases I have mentioned are but the tip of the iceberg. Cloned human embryonic stem cells have the potential to cure, or assist in the discovery of a cure, for just about any degenerative disease or disorder you can imagine. Pliable undifferentiated cloned cells, genetically identical to you, your salvation from the diseases that ail you. Therapeutic human cloning truly is a modern day miracle." (Waite 2001)

Vertebrate cloning timeline

1885   *Weismannn* proposes the theory that there is a loss of genetic information (genes) as cells specialise; this theory later proves to be incorrect.

1914   *Spemann* performs first nuclear transfer with early newt embryonic nuclei.

1938   *Spemann* publishes *Embryonic Development and Induction*, in which he suggests a "fantastical" future experiment: the cloning of an adult animal.

1952   *Briggs* and *King* perform first nuclear transfer where early embryonic Rana frog nuclei are microinjected into enucleated eggs (that is, eggs with their DNA removed).

1958   *Fischberg, Elsdale* and *Gurdon* clone sexually mature *Xenopus* frogs using late embryonic nuclei.

1962   *Gurdon* uses differentiated donor nuclei to clone sexually mature *Xenopus* frogs.

1975   *Bromhall* performs mammalian nuclear transfer in rabbits, but clones do not develop past the embryonic stages.

1979   *Willadsen* splits early sheep embryos and artificially produces "identical" (monozygotic) twins.

1981   *Illmensee* and *Hoppe* claim to have cloned three adult mice by nuclear transfer, but their result has never been repeated.

1983   *McGrath* and *Solter* perform mouse nuclear transfer, but only between two zygotic cells, which are basically just fertilised eggs.

1986   *Willadsen* successfully clones sheep from embryonic (undifferentiated) cell nuclei.

1993   *Hall* artificially clones humans via "embryo splitting", but does not implant the embryos. Embryo splitting is what happens naturally when identical twins are born.

1995   *Campbell* and *Wilmut* clone sheep from differentiated cell nuclei (announced in 1996).

1996   *Wilmut* clones Dolly the sheep using adult cell nuclei (announced in 1997).

1997   *Meng* and *Wolf* clone monkeys, but only from embryonic cell nuclei.

49

1998	*Wakayama* clones fifty mice from adult cell nuclei. This is the first conclusive proof that Dolly was not a fluke.
1999	*Bagushi* clones goats from foetal tissue and *Wells* manages to clone cows from adult cell nuclei.
2000	Three groups succeed in cloning pigs.
2001	*Zavos* and *Antinori* propose to reproductively clone humans.

References

American Society of Reproductive Medicine (ASRM) (2000). "Embryo splitting for fertility treatment." http://www.asrm.com/Media/Ethics/embsplit.html

Bagushi, A. *et al.* (1999). "Production of goats by somatic cell nuclear transfer." *Nature Biotechnology.* 17, 456-461.

BBC News (2001). "UK enters the clone age." http://news.bbc.co.uk/hi/english/uk_politics/newsid_1132000/1132034.stm

Betthauser, J. *et al.* (2000). "Production of cloned pigs from *in vitro* systems." *Nature Biotechnology.* 18, 1055-1059.

Briggs, R. and King, T.J. (1952). "Transplantation of living nuclei from blastula cells into enucleated frogs' eggs." *Proc. Natnl. Acad. Sci. USA.* 38, 455-463.

Bromhall, J.D. (1975). "Nuclear transplantation in the rabbit egg." *Nature.* 258, 719-722.

Brower, V. (2001). "Does imprinting make healthier human clones?" http://news.bmn.com

Campbell, K.H.S., McWhir, J., Richie, W.A. and Wilmut, I. (1996). "Sheep cloned by nuclear transfer from a cultured cell line." *Nature.* 380, 64-66.

Cibelli, J.B., Stice, S.L., Golueke, P.J., Kane, J.J., Jerry, J., Blackwell, C., Ponce de Leon, F.A., Robl, J.M. (1998). "Cloned transgenic calves produced from nonquiescent foetal fibroblasts." *Science.* 280, 1256-1258.

Dinnyes, A. *et al.* (2001). "Development of cloned embryos from adult rabbit fibroblasts: effect of activation treatment and donor preparation." *Biol. Reproduction.* 64, 257-263.

Fischberg, M., Gurdon, J.B. and Elsdale, T.R. (1958). "*Nuclear transplantation in Xenopus laevis.*" *Nature, Lond.* 181, 424.

Gurdon, J.B. (1961). "The transplantation of nuclei between two subspecies of *Xenopus laevis.*" *Heredity.* 16, 305-315.

Gurdon, J.B. (1962). "Adult frogs derived from the nuclei of single somatic cells." *Dev. Biol.* 4, 256-273.

Gurdon, J.B. (1986). "Nuclear transplantation in eggs and oocytes." *J. Cell Sci. Suppl.* 4, 287-318.

Gurdon, J.B. and Colman, A. (1999). "The future of cloning." *Nature.* 402, 743-746.

Gurdon, J.B. and Laskey, R.A. (1970). "The transplantation of nuclei from single cultured cells into enucleate frogs' eggs." *J. Embryol. exp. Morph.* 24, 227-248.

Gurdon, J.B., Laskey, R.A. and Reeves, O.R. (1975). "The developmental capacity of nuclei transplanted from keratinized skin cells of adult frogs." *J. Embryol. Exp. Morph.* 34, 93-112.

Hall, J.L., Engel, D., Gindoff, P.R., *et al.* (1993). "Experimental Cloning of Human Polyploid Embryos Using an Artificial Zona Pellucida." *The American Fertility Society* conjointly with the *Canadian Fertility and Andrology Society,* Program supplement, 1993 *Abstracts of the Scientific Oral and Poster Sessions,* Abstract 0-001, S1.

Howlett, S.K., Barton, S.C. and Surani, M.A. (1987). "Nuclear cytoplasmic interactions following nuclear transplantation in mouse embryos." *Development.* 101, 915-923.

Illmensee, K. and Hoppe, P. (1981). "Nuclear transplantation in *Mus muscularus:* Developmental potential of nuclei from preimplantation embryos." *Cell.* 23, 9-18.

Jaenisch, R. and Wilmut, I. (2001). "Don't clone humans!" *Science.* 292(5517):639.

Killian, J.K. *et al.* (2001). "Divergent evolution in M6P/IGF2R imprinting from the Jurassic to the Quaternary." *Human Molecular Genetics.* 10, 1721-1728.

Kind, A. and Colman, A. (1999). "Therapeutic cloning: needs and prospects." *Semin. Cell Dev. Biol.* 10(3):279-286.

Meng, L. Ely, J.J., Stouffer, R. L. and Wolf, D.P. (1997). "Rhesus monkeys produced by nuclear transfer." *Biol. Reproduction.* 57, 454-459.

McGrath, J. and Solter, D. (1984). "Inability of mouse blastomere nuclei transferred to enucleated zygotes to support development *in vitro.*" *Science.* 266, 1317-1319.

McLaren, A. (2000). "Cloning: Pathways to a pluripotent future." *Science.* 288(5472): 1775.

Mitalipov, S.M. and Wolf, D.P. (2000). "Mammalian cloning: possibilities and threats." *Ann. Med.* 32(7):462-468.

National Academy of Sciences (2001). "Scientific and medical aspects of human cloning." http://www.reproductivecloning.net/nas.htm

Onishi, A., Iwamoto, M., Akita, T., Mikawa, S., Takeda, K., Awata, T. and Hanada Hm Perry, A. C. (2000). "Pig cloning by micro-injection of foetal fibroblast nuclei." *Science.* 289(5482):1188-1190.

Pence, G.E. (1998). *Flesh of my flesh: The ethics of cloning humans.* Rowman and Littlefield Publishing, Inc. Maryland. ISBN: 0847689824.

Platt, J.L. (1999). "Genetic therapies and xenotransplantation." *Expert Opin. Investig. Drugs.* 8(10):1653-1662.

Polejaeva, I.A., Chen, S.H., Vaught, T.D., Page, R.L., Mullins, J., Ball, S., Dai, Y., Boone, J., Walker, S., Ayares, D.L., Colman, A. and Campbell, K.H.S. (2000). "Cloned pigs produced by nuclear transfer from adult somatic cells." *Nature.* 407, 86-90.

Schnieke, A.E. *et al.* (1997). "Human factor IX transgenic sheep produced by transfer of nuclei from transfected foetal fibroblasts." *Science.* 278(5346):2038-2039.

Simnet, J.D. (1964). "The development of embryos derived from the transplantation of neural ectoderm cell nuclei in *Xenopus laevis.*" *Dev. Bio.* 10, 467-486.

Society for the Protection of Unborn Children (2000) "Arguments against UK government human cloning proposals." http://www.spuc.org.uk/cloning/mpbrief.htm

Spemann, H. (1938). *Embryonic Development and Induction.* New Haven, Conn: Yale University Press.

Turner, R.C. (1997). *Human Cloning: Religious Responses.* John Knox Publishing, London. ISBN: 0664257712.

Waite, G. (2001). "Cloning, stem cells and infertility."
http://www.reproductivecloning.net

Wakayama, T., Tateno, H., Mombaerts, P. and Yanagimachi, R. (2000). "Nuclear transfer into mouse zygotes." *Nature Genetics.* 24, 108-109.

Weismann, A. (1892). *Das Keinplasma. Eine Theorie der Vererbung.* Jena: G. Fischer. (Translation in English, 1893, Walter Scott Publishing, London)

Wells, D.N., Misica, P.M., Day, A.M. (1999). "Production of cloned calves following nuclear transfer with cultured adult mural granulosa cells." *Biol. Reproduction.* 60, 996-1005.

Westhusin, D. (1998). "Missyplicity project."
http://www.missyplicity.com

Willadsen, S.M. (1979). "A method for culture of micromanipulated sheep embryos and its use to produce monozygotic twins." *Nature, Lond.* 277, 298-300.

Willadsen, S.M. (1981). "The developmental capacity of blastomeres from 4 and 8 cell sheep embryos." *J. Embryol. exp. Morph.* 65, 165-172.

Willadsen, S.M. (1986). "Nuclear transplantation in sheep embryos." *Nature.* 320, 63-65.

Wilmut, I., Campbell, K. and Tudge, C. (2000). *The Second Creation; The age of biological control by the scientists who cloned Dolly.* Headline Book Publishing, London. ISBN: 0 7472 7530 0

Wilmut, I., Schnieke, A.E., McWhir, J., Kind, A. J. and Campbell, K.H.S. (1997). "Viable offspring derived from foetal and adult mammalian cells." *Nature.* 385, 810-813.

Young, L.E. *et al.* (2001). "Epigenetic change in IGF2R is associated with foetal overgrowth after sheep embryo culture." *Nature Genetics.* 27(2):153-154.

Zavos, P. (2001). "Testimony before the House Subcommittee on Oversight and Investigation; Hearing on Issues Raised by Human Cloning Research."
http://www.reproductivecloning.net/articles/testimony.htm

Cloning Dolly

by Professor Keith H. S. Campbell

Introducing Dolly – background to her creation

Animal reproduction

In animals, reproduction occurs by sexual means; fertilisation of the female egg by the male-derived sperm results in the production of a single cell or zygote*, which begins development and results in the production of offspring. The genome, or genetic information present in the majority of the cells of the body, consists of two sets of genes, one contributed by the sperm (the paternal genome) and the other by the egg (the maternal genome).

During the development of egg and sperm cells the genetic information is rearranged by the process of meiosis*; thus each fertilised zygote contains a unique genome and results in the formation of a unique individual. During development the zygote has to grow and divide. With each division the genome is copied and each cell inherits a copy of this novel genome that is located within the nucleus (also referred to as nuclear or chromosomal DNA).

Under natural circumstances the occurrence of "clones" is restricted to the production of identical twins due to the division of a single embryo to form two identical but separate individuals (also termed "monozygotic", derived from a single zygote). In some species division of the embryo may result in the formation of more than two clones (for example the armadillo). Such "splitting" of embryos may also be carried out experimentally in the laboratory to produce identical offspring.

Nuclear transfer

In contrast to the "splitting" of embryos, Dolly was produced by a technique known as nuclear transfer.

In this technique genetic material (the nucleus) is usually removed from an unfertilised egg (recipient cell) and replaced with the nucleus of a cell at a later developmental stage (donor cell). A pulse of electricity is used to fuse the new egg and the

Zygote: the cell resulting from the fusion of an egg and a sperm (i.e. a fertilised egg).

Meiosis: the process by which germ cells (i.e. those in the ovaries or testes) divide to produce gametes. In meiosis I, homologous chromosomes exchange genetic material. In meiosis II the two resulting diploid cells (i.e. which contain two sets of chromosomes) with their recombined chromosomes divide further to form two haploid gametes (i.e. which contain only one set of chromosomes).

55

Mitochondria:
energy-producing bodies in the cell, which contain a small genome.

Organelle:
specialised part of a cell.

new nucleus. No fertilisation occurs and therefore the resulting zygotic nuclear DNA is not derived from a maternal and a paternal genome.

In normal sexual reproduction the egg and the sperm are equal contributors of nuclear DNA, but the egg contains numerous factors essential for development. In particular the egg contains extrachromosomal DNA located in the mitochondria*; these intracellular organelles* are inherited primarily through the maternal line. In the true sense of the word, then, the animals produced by nuclear transfer are not "clones", as eggs used in the process are obtained from different females and therefore maternally inherited factors (that is, mitochondria) will differ between the resultant offspring. Although the offspring are generated by asexual means, cell duplication or splitting did not occur and therefore they may more aptly be described as "genomic copies".

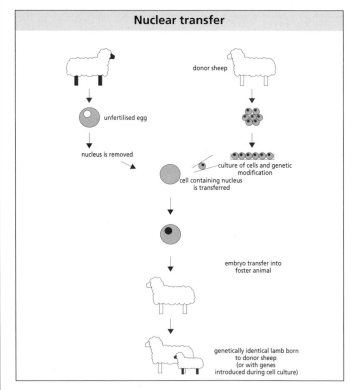

Factors which dictate the success of the nuclear transfer experiments

A wide range of factors dictate the success of nuclear transfer experiments, including:

- the quality of the recipient cell;
- the quality of the donor cell;
- the cell cycle phases, which can be broadly divided into cell division (meiosis), and interphase (the main part of the cycle of the cell, between periods of division);
- the differentiated state of the donor cell;
- how far along the differentiation pathway we are;
- chemicals and media being used.

After fusion, reconstructed eggs are grown in culture for about seven days and only then are the ones that might have a chance of going on to develop to term transferred to a pseudo-pregnant recipient. During the experiments leading up to the production of Dolly, 277 reconstructed eggs (each with a diploid nucleus from the adult animal) were cultured until day 7. Twenty-nine of the eggs that appeared to have developed normally to the blastocyst* stage (that is, the stage at which an embryo is a hollow ball of cells) were implanted into thirteen surrogate ewes and only one gave rise to a live lamb.

Exceptional factors in the production of Dolly

Dolly's cloning was an extremely "newsworthy" event. However, apart from the technique used (cloning), what also makes the birth of Dolly exceptional from a biological perspective is that it dispelled the previously held belief that cell differentiation was invariable, that is, a differentiated cell could not recover its pluripotency*.

Doubting Dolly

There have been doubts from some quarters about the origin of the donor cell used to create Dolly. In a letter to *Science* magazine in 1998, doubts about the process which led to Dolly's birth and the observation that she was only one out of 277 reconstructed eggs to make it to term were raised by

Blastocyst: the hollow sphere of cells that develops from the morula (the solid mass of cells produced by the first divisions of a fertilised egg) and implants in the uterine wall.

Pluripotency: possessing the capacity to differentiate along a variety of pathways.

Norton Zinder of the Rockefeller University of New York and Vittorio Sgaramella of the University of Calabria in Italy. It was notably pointed out that the ewe which supplied the cell that made Dolly was in the last stage of pregnancy at the time of her death; therefore a stray foetal cell in the blood stream of the mother ewe might have caused Dolly to survive, rather than a purely adult cell.

These doubts were disproved scientifically twice through carrying out extremely sensitive genetic fingerprinting* (both immediately after Dolly was born and after the aforementioned letter). Indeed the proportion of circulating foetal cells is between 1 in 100 000 and 1 in a billion and so the likelihood that a stray foetal cell was responsible is highly improbable. More recently, research carried out by Jean-Paul Renard on calves at INRA[1] in Paris has also shown that differentiated cells can give rise to cloning, which supports the argument that Dolly is not just a one-off. Every day more and more examples can be given. Since completing experiments on sheep, other animals have also been cloned using nuclear transfer, includ-ing mice, pigs, goats and cats.

Dolly is not, however, an identical twin to her cell donor mother, because her cytoplasm came from a Scottish Blackface ewe, despite the fact that it was a 6-year-old Finn Dorset which provided the mammary glands. DNA clones are not necessarily identical as it is not just genetic makeup which dictates the appearance of the clone. Cloning a beloved family pet might be a disappointing exercise, for example, because the resultant animal might be slightly different in appearance, even though it had the same DNA. This is true of the first domestic animal formed in the same way as Dolly, namely "Carbon copy" the first cloned kitten.[2]

Problems and abnormalities

The only true measure of the efficiency of the nuclear transfer process is the production of viable offspring. The development of reconstructed embryos is influenced by many factors as has already been discussed. Similarly, induction and maintenance of pregnancy is dependent upon a range of factors influenced both by the quality of the transferred embryo and the age, seasonality, nutritional and hormonal status of the surrogate recipient.

The development of offspring using nuclear-transfer reconstructed embryos is an inefficient process with the majority of studies reporting between 0.5 and 5.0% development to term. In the case of Dolly, for example, she was the only one to survive out of 277 reconstructed mammary cells. Losses occur throughout gestation, at birth, and following birth, and a range of abnormalities have been reported, including excessive weight, premature ageing and in the case of Dolly, arthritis.[1] The reasons for these abnormalities are unknown but may reflect incomplete or inappropriate reprogramming, possibly related to problems with imprinted genes*. A greater understanding of normal development control may help elucidate the mechanisms involved in these processes.

Imprinted genes: copy of genes derived from either the maternal or paternal genome that are suppressed in the embryo.

Animal reproductive cloning

White sheep

Black-faced sheep

Unfertilised egg

Adult cells

Embryo

Throw away genes

Embryo

Throw away empty cell

Embryo implanted into womb of sheep

• Black faced
• Genetically identical to the sheep who provided adult cells

Dolly born

1.
Article on Dolly and arthritis :http ://news. b b c . c o . u k / h i / english/sci/tech/news id_1741000/1741559 .stm

Blastomere:
the cells that make up the zygote.

Integration:
the incorporation of an injected gene into the genome of an organism.

Mosaic embryo:
embryos in which some cells contain a copy of the injected gene and others do not.

Transgenic embryo:
an embryo that contains a gene transferred from one organism into another.

Germ line:
any cell in the series of cells that eventually produce sperm and eggs.

Phenotype:
influence of the environment.

Future prospects

Agriculture

Nuclear transfer using embryonic blastomeres* as nuclear donors has a number of applications in agriculture and research for multiplication of élite embryos or for the production of multiple copies for research purposes. However, these applications are limited by the number of donor cells available and the efficiency of the process. The use of cultured cell populations can increase the number of animals which may be produced from an élite embryo, foetus or adult. In addition, the storage of frozen cell populations may prove useful in the preservation of genetic resources in a number of species. However, in the short term, the major implication of the use of cell populations that may be maintained in culture prior to their use as nuclear donors is the provision of a route for the precise genetic modification of farm animal species.

Production of a transgenic animal can be achieved by the injection of the required gene into the pronucleus of a zygote. Although this technique has been applied successfully in a number of species including mice, rabbits, pigs, sheep, goats, cattle (for review see Wall *et al.* 1992), there are a number of disadvantages:

- integration* does not always occur during the first cell cycle resulting in the production of mosaic embryos* (Burdon *et al.* 1992);

- integration occurs at random within the genome, resulting in variable expression of the gene product. At present only simple gene additions may be performed;

- the selection of transgenic* embryos prior to their transfer is hampered by mosaicism (Rusconi 1991);

- the production of the required phenotype* coupled to germ line* transmission may require the generation of several transgenic lines;

- multiplication of the required phenotype or its dissemination into the population is restricted by breeding programmes.

In contrast, the production of offspring from a single cell, or cloned population offers significant advantages. Genetic modification can be performed in culture and the modified cells selected prior to animal production. It will be possible to remove (knockout) as well as to add genes; also, precise modification of control regions or addition of genes to specific regions of the genome (knockin) will be facilitated. The production of an animal from a single nucleus removes the problems associated with mosaicism as all of the cells within the resultant animal will contain the modification which will be transmitted through the germ line. All of the animals produced will be transgenic and flock or herd generation can be accelerated by producing multiple copies from the cultured cells. The experiments that led to Dolly involved the use of mammary epithelial cells. The use of this cell line was related to the possible screening of transgenic cells for milk production potential *in vitro* prior to animal production. Thus it may be possible to predict expression level and select the highest expressing cell populations prior to animal production.

In order to carry out these modifications the cultured cell populations must be amenable to transfection* and selection in culture and maintain their ability to be used for successful nuclear transfer. It has been demonstrated that foetal fibroblasts* are suitable for this purpose with the production of Polly, a nuclear transfer lamb transgenic for human factor IX derived from a transfected, selected cell population (Schnieke *et al.* 1997).

These experiments also demonstrated that the efficiency of animal production was increased over twofold, in terms of total animals used, as compared to pronuclear injection. The generation of animals carrying multiple genetic modifications requires the sequential addition, removal or modification of specific genes. In culture, primary cell populations have a finite lifespan; however, by re-deriving cell populations from embryos, foetuses or offspring produced by nuclear transfer it will be possible to extend the period in which cells can be maintained in culture to carry out these modifications. Genetic modification may also lead us to produce disease-resistant animals and allow us to modify production traits, for example,

Transfection: the act of "infecting" a cell with DNA to effect gene transfer.

Fibroblast: fibrous connective tissue cells that are flattened and irregularly shaped.

Endogenous gene: one that comes from inside the body.

Cystic fibrosis: a lung disease that causes the production of thick mucus in the lungs, hampering breathing.

modification of meat quality in a range of species or wool quality in sheep.

Human medicine

The ability to carry out precise genetic modification on cells in culture not only provides a method for the improvement of present transgenic technology but also facilitates previously improbable genetic modifications. Transgenic animals can play a role in a range of human therapies; these will be discussed in relation to present and future therapies.

Biopharmaceuticals: the production of human proteins in transgenic animals. Human proteins may be produced in a range of tissues and bodily fluids including blood, urine and milk. Although each of these may play a particular role the value of biopharmaceutical production in transgenic animals lies in the high volume that can potentially be produced at relatively low cost. To this end the production of proteins in the milk of sheep, goats and cattle provides a useful route, although for products required in small amounts, transgenic rabbits may also be used. A range of therapeutic proteins is being produced in the milk of transgenic animals including alpha-1-antitrypsin (for treatment of cystic fibrosis*) and Factor IX (for treatment of haemophilia B) (for review see Garner *et al.* 1998). Nuclear transfer will facilitate the removal of endogenous genes* to aid purification, for example the replacement of bovine serum albumin with human serum albumin (HSA) in order to produce large amounts of HSA for the treatment of burns.

Nutraceuticals: the modification of animal milk to enhance nutritional value, removal of allergens (for example, beta-lacto-globulin in cattle).

Xenotransplantation: the use of animal organs and other tissues for human transplantation. Physiologically, pig organs are similar to humans and are considered suitable for transplantation. However, there is a major problem with organ rejection. Although all of the mechanisms that are involved in this rejection are not completely understood, it is known that a major antigen involved is alpha-1,3 galactose. This is present on the surface of pig cells but is not found in humans who therefore mount an immune response. Nuclear transfer from cultured cells will facilitate knockout of the pig gene coding for alpha-1,3 galactosyl transferase. Potential organs and tissues for transplantation include the heart, lungs, kidneys and pancreatic islets (treatment of diabetes) (for review see White, 1998).

Disease models: animal models which are mostly available for the study of human genetic disorders (at this time mostly mice). However, this species may not manifest the same clinical symptoms as humans. Nuclear transfer technology will allow the production of disease models in species that are physiologically more similar to the human in order to follow disease progression or to assess the benefits of any potential new therapies (including gene therapy). An example of this is cystic fibrosis that manifests in the gut and not the chest in mice.

Ageing: The study of this is another application of the nuclear transfer techniques that is currently being researched.

Preservation of genetic diversity: This can occur by cryopreservation (freezing) of cells from rare or endangered species. This may be of particular use in rare livestock species.

Examples with indirect effects on human health

Genetic modification of farm animals may be used to improve various production or nutritional traits. In addition improvement of animal health by the introduction of genes for disease resistance or removal of genes for disease susceptibility (for example, PrP gene involved in scrapie and BSE) may have long-term implications for human health.

Role of nuclear transfer in stem cell therapies

Nuclear transfer provides a route to the de-differentiation of somatic cells, providing embryonic cells that may form the basis for isolation of stem cell populations.

Thoughts on human cloning

To conclude I would like to mention my particular opinion about human cloning. Since the creation of Dolly this possibility has captivated the attention of the media. Despite biological differences between humans and sheep, it should be possible to clone humans. I think it is highly questionable, however, that if we did clone people we would succeed in producing the same person. It depends on the extent to which one thinks genes are responsible for personality and behaviour. While there is some genetic component to personality, I do not think that this is the "be all and end all". Everything depends on interaction with environmental, familial, socio-economic, cultural and other factors. Cloning DNA from Adolf Hitler might not lead to people who behave in a similar way. One concern is that cloning is not a very efficient process and a certain number of abnormalities may result from this. Nor is it known how these clones will survive in the long term. As has been mentioned in other chapters there would be significant pressure on the child to behave like the person they were cloned from.

Possible uses of cloning humans would be to remove genetic defects from embryos, although in the UK pre-embryo genetic screening (taking cells from embryos and checking to see whether or not they have diseases) is already available.

A possible useful application of human cloning, which is not necessarily reproductive cloning as it would not involve cloning an individual, would be if one were to take a zygote (that is, the product of a sperm and an oocyte and thus from sexual reproduction) with a genetic defect (for example, muscular dystrophy), make a stem cell line from the resulting embryo, remove the genetic defect, clone the corrected stem cell by nuclear transfer, make the embryo, transplant it back into the mother, and once a pregnancy is established, dispose of the rest of the cultured stem cells. The result would be a unique individual baby, which is the product of sexual reproduction, but whose genome is corrected, including in the germ line, so that it no longer has a genetic defect. However, cloning of humans by nuclear transfer is still not a reality.

References

Briggs, R. *et al.* (1952)."Transplantation of living nuclei from blastula cells into enucleated frog's eggs", *Proc. Natl. Acad. Sci. USA*, 38, p. 455

Burdon, T.G. *et al.* (1992) "Fate of microinjected genes in pre-implantation mouse embryos", *Mol. Rep. Dev.*, 33, p. 436

Campbell, K.H.S. *et al.* (1994)."Recent advances on *in vitro* culture and cloning of ungulate embryos", *Fifth World Congress on genetics as applied to livestock*, p. 180

Campbell, K.H.S. *et al.* (1996a). "Cell cycle co-ordination in embryo cloning by nuclear transfer", *Rev. Reprod.*, 1, p. 40

Campbell, K.H.S. *et al.* (1996b). "Sheep cloned by nuclear transfer from a cultured cell line [see comments]", *Nature*, 380, p. 64

Campbell, K.H.S. *et al.* (1993). "Nuclear-cytoplasmic interactions during the first cell cycle of nuclear transfer reconstructed bovine embryos: implications for deoxyribonucleic acid replication and development", *Biol. Reproduction*, 49, p. 933

Campbell, K.H.S. *et al.* (1997). "Totipotency or multipotentiality of cultured cells: applications and progress", *Theriogenology*, 47, p. 72

Campbell, K.H.S. *et al.* (1998). "Nuclear transfer in animal breeding": In A.J Clark (ed.), *Technology for the 21st Century*, Harwood Academic Publishers, Amsterdam, p. 47

Cheong, H.T. *et al.* (1993). "Birth of mice after transplantation of early cell-cycle-stage embryonic nuclei into enucleated oocytes", *Biol. Reproduction*, 48, p. 958

Czolowska, R. *et al.* (1984). "Behaviour of thymocyte nuclei in non-activated and activated mouse oocytes", *J. Cell. Sci.*, 69, p. 19

Di Beradino, M.A. (1997). *Genomic potential of differentiated cells*, Columbia University Press, Columbia.

Driesch, H. (1892) "Entwicklungsmechanische Studien. I. der Wert der beiden ersten furchungszellen in der Echinoderenenttwicklung. Experimentelle Erzeugug von teilund doppelbildungen.", *Zeitschr. Wiss. Zool*, 53, p. 160

Garner, I. *et al.* (1988). "Therapeutic proteins from livestock in animal breeding": In A.J. Clark (ed) *Technology for the 21st Century,*. Harwood Academic Publishers, Amsterdam, p. 215

Gurdon, J.B. *et al.* (1966). "Fertile intestine nuclei", *Nature*, 210, p. 1240

Gurdon, J.B. *et al.* (1975). "The developmental capacity of nuclei transplanted from keratinized skin cells of adult frogs", *J. Embryol. exp. Morphol.*, 34, p. 93

McCreath, K.J. *et al.* (2000). "Production of gene-targeted sheep by nuclear transfer from cultured somatic cells", *Nature*, 405, p. 1066

McGrath, J. *et al.* (1993). "Nuclear transplantation in mouse embryos", *J Exp. Zool.*, 228, p. 355

McGrath, J. *et al.* (1998). "Nuclear transplantation in the mouse embryo by microsurgery and cell fusion", *Science*, 220, p. 1300

Prather, R.S. *et al.* (1989). "Nuclear transplantation in early pig embryos", *Biol. Reproduction*, 64, p. 414

Robl, J.M. *et al.* (1987). "Nuclear transplantation in bovine embryos", *J. Anim. Sci.*, 64, p. 642

Roux, W. (1888). "Beitrage zur Entwickelungsmechanik des Embryo. ueber die kunstliche hervorbringung halber embryonen durch Nachtwickelung einer der beiden ersten Furchungskugeln, sowie uber die Nachtwickelung (Postergeneration) der Fehlenden Korperhalfte." *Virchows Arch. Anat. Physiol.*, 21, p. 113

Rusconi, S. (1991). "Transgenic regulation in laboratory animals" *Experientia*, 47, p. 866

Schnieke, A.E. *et al.* (1997). "Human Factor IX transgenic sheep produced by transfer of nuclei from transfected foetal fibroblasts", *Science*, 278, p. 2130

Spemann, H. (1938). *Embryonic development and induction,* Garland Publishing Inc., New York, p. 210.

Wakayama, T. *et al.* (1998a). "Full-term development of mice from enucleated oocytes injected with cumulus cell nuclei", *Nature,* p. 369

Wall, R.J. *et al.* (1992). "Transgenic farm animals – A critical analysis", *Theriogenology,* p. 337

Weismann, A. (1892) *Das kleimplasma. Eine theorie de Verebung.*

White, D. *et al.* (1998). "Xenografts from livestock in animal breeding": In A.J. Clark (ed.) *Technology for the 21st Century.* Harwood Academic Publishers, Amsterdam, p. 229

Willadsen, S.M. (1986) "Nuclear transplantation in sheep embryos", *Nature,* 320, p. 63

How human reproductive cloning could change our lives – some scenarios

by Professor Claude Sureau

The term "cloning" has already been discussed on several occasions in previous chapters; suffice it to say that it covers a wide range of situations, in terms of the beings or things to be cloned (vegetable, animal, fragments of genome), the techniques used and, most importantly, the motives on which it is based or likely to be based, particularly in the case of human cloning.

We will consider in turn a number of specific examples designed to explain the biological and clinical phenomena in question: some human scenarios that are patently indefensible, others that merit more detailed consideration and, finally, some highly specific areas, the study of which may help temper some of the dogma that surrounds this subject.

Specific examples to explain the phenomena in question

We will begin by calling to mind the terms of the "Additional Protocol to the Convention for the Protection of Human Rights and Dignity of the Human Being with regard to the Application of Biology and Medicine, on the Prohibition of Cloning Human Beings", which was opened for signature on 12 January 1998 and entered into force on 1 March 2001:[1]

"Article 1:

1. Any intervention seeking to create a human being genetically identical to another human being, whether living or dead, is prohibited.

2. For the purpose of this article, the term human being 'genetically identical' to another human being means a human being sharing with another the same nuclear gene set."

Before turning our attention to humans, let us consider some well-known examples in the plant and animal kingdoms.

1.
Short title for the parent convention is the Convention on Human Rights and Biomedicine. See : http://book.coe.fr/conv/en/ui/ctil/menu-en.htm.
See also Appendix III.

Epigenesis:
all factors that influence the expression of the genotype of an organism during its development, e.g. environmental factors.

Fission:
splitting or division.

Daughter cell:
progeny of a cell arising from a mitotic division (identical to a parent cell).

Example from the plant kingdom: propagation by cutting

This example, incidentally, matches the original definition of cloning, in that it involves the reproduction of a plant by "vegetative multiplication", that is, from a non-sexually differentiated part of the organism in question. Put simply, if you cut off a branch from a poplar tree and plant it, it will produce a new poplar which will be a clone of the first, in that its cells will contain in their nuclei the same gene set as that found in the cells of the original tree. The new poplar, however, will not necessarily be identical to the original; indeed, there is even a good chance that it will be very different, firstly because its biological age will not be the same, and secondly, and most importantly, because its development will be influenced by the environment: depending on the soil quality, climatic variations and the influence of surrounding trees, the growth and appearance, what biologists call the "phenotype" (as opposed to the genotype, that is, the genetic make-up), will be different from those of the original tree. Everything that thus serves to modulate the expression of the genotype, whether through environmental impact on development, or even, most importantly, through direct interference with the functioning of the genome, constitutes epigenesis*. An important initial observation can thus be made: because of epigenesis, two genetically identical individuals remain fundamentally distinct; the one can never be a "carbon copy" of the other.

Example from the animal kingdom: the amoeba

A primitive yet extremely common, animal life form, the amoeba, reproduces by binary fission*, that is, amoebae do not use a specialised reproduction system either, but simply split into two daughter cells*; genetically identical to the original organism. These cells will then develop according to their fate, once again modulated by the environment.

This is an interesting example because it could be argued that, barring fatal accidents, amoebae are immortal. Immortality thus arises from the ability to divide into two daughter cells, and not only the ability to generate another genetically identical individual. We will return to this later.

Monozygotic twinning in viviparous species**

A brief reminder, firstly, about the process known as fertilisation: the spermatozoa swim to the egg, or oocyte, which is transported along the fallopian tube after being caught by the infundibulum*. The spermatozoa surround the oocyte and cling to the surface. Eventually, one of them manages to penetrate. Is this, then, when fertilisation occurs? The moment in time that defines the beginning of a new being, and with it the protection that is accorded to him or her by the law in many countries, save in exceptional circumstances?

The answer is far from certain, because a full thirty hours will elapse from the time the sperm manages to penetrate the egg to the point at which the membranes surrounding these two pronuclei* break down and the two sets of chromosomes mingle and pair off, thereby restoring the diploid* chromosome number 46 (each gamete nucleus having only 23 chromosomes). During this time, the sperm must cross the various natural barriers of the egg, travel through the egg's cytoplasm and approach its nucleus – without, contrary to popular belief, fusion occurring to produce a single cell with a single nucleus (the mythical original cell postulated by many theoreticians). Concurrent with pronuclei break down, the cytoplasm begins and completes its division (so that there is a seamless progression from the "fertilised" egg stage with its two pronuclei, to the 2-cell embryo stage). The time required for this slow "fertilisation" is thus thirty hours or more, a somewhat inconvenient fact as it makes it difficult to determine the precise moment in time when life begins, or rather a being, especially a human being, comes into existence. This is no small matter, for it will be appreciated that any action liable to interfere with this biological development and to bring it to a halt may, depending on how one sees it, be classed as abortion or contraception. Unless, conversely, the recognition of these phenomena and their complex nature leads to a more subtle exploration of the meaning of the words and their conceptual significance.

There are, in any case, other questions which need to be considered: can "conception" be deemed to have taken place once the 2-cell "zygote*" is formed? Or must we wait another few days until the new being's genome has fully expressed itself,

Monozygotic twinning:
the production of two embryos from the splitting of a single zygote (so-called "identical" twins).

Viviparous species:
a species that gives birth to live young rather than laying eggs. Includes all mammals except monotremes.

Infundibulum:
the funnel shaped part of the fallopian tubes into which the egg is drawn after ovulation.

Pronuclei:
the haploid nuclei of the egg and sperm after fertilisation but before they fuse and undergo the first mitotic division.

Diploid:
containing a paired set of each chromosome (46 in humans). All somatic cells are diploid.

Zygote:
the cell resulting from the fusion of an egg and a sperm (i.e. a fertilised egg).

Morula:
the solid mass of cells produced by the first cell divisions of a fertilised egg. The stage preceding the blastocyst.

Blastocyst:
the hollow sphere of cells that develops from the morula (the solid mass of cells produced by the first divisions of a fertilised egg) and implants in the uterine wall.

Dizygotic twins:
twins derived from two separate zygotes, which have been fertilised by a separate sperm (so-called "non-identical" twins).

independently from those of each of the original gametes, whose influence (parental imprint) will gradually fade. Surely it is only then that a new "individual", biologically and genetically unique, can be said to have come into existence?

The very term "individual" is itself open to question, if only etymologically, as it may divide further right up to day 14. In the meantime, by day 7, it will, with luck, have implanted itself in the mucous membrane of the uterus, thus marking the start of the pregnancy proper. Should we, as some contend, allow this implantation to define the start of the symbolic existence of the new being, by making this implantation the basic condition for conferral of the status of *infans conceptus*, as accorded by French law? Once again, the question is an important one, because depending on how we answer it, certain actions (whether hormonal or mechanical) will be construed as contraception or abortion.

Between day 2 and day 14, then, this embryo (the morula* and later the blastocyst*) may divide into "twins" who will have the same genetic constitution and will therefore be classed as identical or monozygotic twins (as opposed to fraternal or dizygotic twins*, which are derived from the fertilisation of two eggs by two sperm, and are no more genetically alike than non-twin siblings. The incidence of dizygotic twins is currently on the rise, owing either to the transfer to the mother's uterus of several embryos obtained from *in vitro* fertilisation, or, far more commonly, from simultaneous fertilisation of several eggs following ovarian stimulation).

The important point to note is that, in the case of dizygotic twins, the fertilisation process and its immediate aftermath are, although simultaneous, the same as in individual pregnancies. With monozygotic twins, however, a single egg is fertilised, which then divides into several "individuals", normally two (although sometimes a good many more in certain animal species such as the armadillo).

If separation occurs early on, the twins may be located in two different sacs (like all dizygotic twins; this characteristic cannot be used, therefore, to differentiate them); later on, they may find themselves in the same sac (possibly separated by a thin wall); later still, and fortunately only very rarely, they may remain partially attached by a segment of their bodies, forming what are popularly known as Siamese twins, after the famous conjoined twins from Siam who were exhibited in Barnum's Circus.

What is interesting, and rightly emphasised in the Additional Protocol to the above-mentioned Convention on Human Rights and Biomedicine,[1] is that this monozygotic twinning is nature's version of one of two basic methods of cloning viviparous species, sometimes referred to as "horizontal" cloning (because the two resulting individuals are the same age) or cloning by embryo splitting, since they are derived from the separation of the original cluster of blastomeres*.

In theory – and this is one initial scenario that could conceivably apply to humans – scientists might envisage carrying out (surgically or medically) artificial monozygotic twinning using zygotes obtained through IVF*. Were they to do so, this would certainly qualify as a form of human cloning.

What rationale might there be for taking this path? It is hard to see one, because generally speaking, embryos thus fertilised *in vitro* tend to be excessively numerous. What would be the point, then, of seeking to obtain monozygotic twins, when dizygotic twins are something which doctors normally do their utmost to avoid? There is no evidence to support the claim that the chances of implantation, for example, in cases where there have been repeated failures, are greater if the twins in question are monozygotic rather than dizygotic. This hypothetical advantage would in any case be negated by the disadvantages associated with twinning.

One possible exception,[2] though, might be made in the case of a childless woman who had had her ovaries removed and was thus incapable of ovulating, but who had previously undergone IVF with preservation of a frozen embryo. Such a woman might be prompted to request blastomere separation, not on some whim, but in order to improve her chances if an initial transfer failed.

Should her request be granted? What grounds might there be for denying it? If we consider the actual terms of the Additional

Blastomere:
the cells that make up the zygote.

IVF:
(*in vitro* fertilisation); biological process that occurs in a container outside of the organism; e.g. cells or tissues grown in culture.

1.
Full title: The Convention for the Protection of Human Rights and Dignity of the Human Being with regard to the Application of Biology and Medicine. See http://book.coe.fr/co nv/en/ui/ctrl/menu-en.htm.

2.
Editor's note:
The opinions expressed in this and the two following paragraphs are not necessarily in agreement with the generally accepted interpretation of the Additional Protocol.

Cryopreserve:
to preserve tissues or organisms by freezing at very low temperatures.

Hemizygotic embryo:
an embryo resulting from half of a zygote.

Nuclear transfer:
the nucleus of a cell from the animal to be cloned is transferred into an oocyte whose own nucleus has been removed. The resulting cell is then cultured to form an embryo that is implanted into a female uterus.

Protocol, there would be no initial cloning involved, but rather blastomere splitting at the zygote stage in order to obtain a pregnancy, and eventually a child, who would be identical only to the other part of the zygote. This would still be frozen, and as yet unconstituted as a "born person".

The Additional Protocol would be infringed only if the two separate parts of the original zygote were implanted at the same time and even then it is hard to see why such an operation designed to obtain two true, genetically identical twins should be deemed unethical when the transfer of two dizygotic embryos, or indeed the birth of natural identical twins, elicits no comment whatsoever.

It could be that the wording of the Additional Protocol is less than ideal in this respect. After all, it is not uncommon for supposedly broad-based and somewhat academic texts of this kind to be out of step with real life.

One would have to question the legitimacy, however, of continuing to cryopreserve* the remaining hemizygotic* embryo, once the first embryo had been born. By rights, the embryo should probably be destroyed, but the mother might argue that she wanted to keep it in case something happened to her first child. In such an event, might there be a case for transferring the remaining twin? If so, how long after the death of the first twin, and should the mother's age be a factor? Wouldn't there be a danger of bypassing the painful yet necessary grieving process? And wouldn't certain parallels begin to emerge between this situation and another that will be examined later, namely nuclear transfer* from a cell of a dying child?

The Additional Protocol, in any case, would not have been observed, as the second child would be genetically identical to its dead sibling.

Likewise, continued cryopreservation of the second embryo while the first twin was still alive would be highly questionable, not only because it would be incompatible with the letter of the Additional Protocol, but also because it would bring us several steps closer to a situation that is fraught with social dangers: the situation of a pair of identical twins, derived from the division of a single zygote and conceived by the same parents, but born at different times, and possibly even carried by different mothers, the first being the genetic mother, and the second a later partner of their common progenitor, twenty or thirty years down the line.

Given the potentially serious social and family consequences associated with supernumerary embryos and hence frozen dizygotic twins, and possibly staggered births, it is surely not advisable to complicate matters further by creating similar situations with identical twins.

The comparison with natural monozygotic twinning also provides us with a vital piece of information, which it will be as well to remember when examining the issue of nuclear transfers: identical, that is genetically identical, twins are not, in spite of appearances, "carbon copies" of each other.

Their morphology is, of course, strikingly similar (except in one very rare instance, known as heterocaryotic monozygotic twinning* where a genetic defect occurs in the very early stages of development, causing partial trisomy 21* in one of the twins, or resulting in monozygotic twins of different apparent sex).

All the studies show, however, that although the main embryological, and in particular morphological, features are the same, epigenesis gives a specific direction to the development of the embryo, and later the foetus, the infant and finally the adult. It particularly influences the nervous system as Salmon (1985) remarked:

"The Utopian notion of cloning a human subject is fundamentally flawed. Genes alone are not enough to define a person. What makes Man what he is and differentiates him from animals and, indeed all other living things, is the extraordinary capacity of his neuronal cells to break free from their genetic programme, a capacity that no other species possesses."

The conclusion to bear in mind for future reference is that naturally occurring monozygotic twinning provides us with a vital, yet often overlooked, clue: two "genetically identical" humans are not really identical after all. It is a fundamental error to equate people with animals. Even the calves cloned by France's INRA[1] have different markings on their hide. Human cloning as represented in its most perfect and natural form by

Heterocaryotic monozygotic twinning:
the production of genetically different twins from the same zygote due to a mutation in one of the nuclei after zygote splitting has occurred.

Partial trisomy 21:
a situation in which there are three copies of chromosome 21, rather than two, but where the third is incomplete. This condition is characterised by some degree of physical and mental abnormalities.

1.
INRA (Institut National de la Recherche Agronomique):
http://www.inra.fr/ENG/

monozygotic twinning shows us that human "procreation" is not simply a matter of "reproduction", and that consequently, observations relating to animals cannot automatically be applied to humans.

Nuclear transfer

The research carried out on animals has already been discussed in detail in previous chapters. Suffice it to say, then, that the first 2 or 4 cells[1] (the blastomeres) of the embryo are what we call "totipotent", that is, each of them possesses in its genome the capacity to generate an entire organism. This accounts for monozygotic twinning, and also pre-implantation diagnosis, to which we will return later.

Broadly speaking, in the early stages of its development, each of the embryo's blastomeres is in itself a "potential embryo". It has only to be separated to become an actual embryo, and implanted to realise its destiny as a "potential person".

Later, as the embryo gradually develops, its cells will specialise, or "differentiate", in a manner consistent with their final destiny (determined by a selection mechanism that is still elusive and little understood) and so "lose" their totipotency. In reality, however, this capability is not lost, but merely masked, in that the genetic mechanisms involved are disabled, or "supressed". Hence the fundamental question that has been posed by scientists for years: is totipotency recoverable? Can cells be "de-differentiated", then "re-differentiated"?

It did not take long to discover that this was possible in embryonic cells, by transferring their nucleus to the cytoplasm of an egg whose own nucleus had been removed, in what is known as nuclear transfer. In the case of embryonic cells, however, which are still very similar to the zygotic cells, this did not come as any great surprise.

1.
The first division of the zygote produces two cells and the second division four cells. These cells are known as blastomeres.

More surprising was the discovery that this capacity existed in more highly developed cells, namely foetal cells. The biggest breakthrough, however, was the discovery, through Dolly, that this de-differentiation could be performed on adult cells, capable, following transfer, of "reactivating" all the capabilities of the original cells, and of re-differentiating them in a manner apparently similar to that of a primitive egg. Transferring the nucleus of an adult cell (somatic cell)* can thus lead to the formation and birth of an individual with the same DNA as that found in the cells of the original individual. This is very definitely a case of what will be referred to as reproductive cloning, by fusion (of the somatic cell nucleus and the egg's cytoplasm)* or vertical cloning (since the individual derived from the cloning and the original belong to different "generations").

Somatic cell: all body cells that are not part of the germ line.

Cytoplasm: all of the living part of a cell within the membrane.

It will be observed that all the of the organism's cells are thus "potential individuals".

There will also arise the question of the nature of the being thus "created": is it an embryo, in that it has the capacity to develop into a foetus and eventually an adult, as the consequentialist, teleological view seems to suggest? Or can it not be classed as an embryo, because it is not derived from the meeting and fusion of two gametes?

In the meantime, though, one point must be emphasised: the success rate of such transfers is remarkably low in the few animal species where it has been shown to work at all: Dolly was born only after 276 failures. Even if the transfer succeeds and embryo-foetal development is completed, the perinatal mortality rate is extremely high, for reasons that are still deeply unclear. Overall, the probability of success is at best between 1 and 4%. There is also still some doubt about the kind of future that awaits such individuals with, for instance, conflicting reports about the development of the telomeres (ends of the chromosomes), which are known to play a part in reducing the lifespan of cells or, on the contrary, lengthening it, resulting in cancer in extreme cases. These are all factors which must naturally be borne in mind when envisaging the application of this technique to humans.

In what circumstances might scientists contemplate such a course of action? These are innumerable and could not possibly all be covered here. Rather than attempting to do so, we will consider them according to the various rationales advanced, some more explicitly than others.

Unrealistic scenarios

Wealthy old megalomaniac's dream of immortality

This was the premise of David Rorvik's remarkable novel (Rorvik 1978), already regarded as a classic. Such a dream is absurd, however, for two reasons. Firstly, unlike the amoeba which lives on in the structure of its "daughters", cloning will not enable the donor of the cell to escape death. Secondly, and this is something we will return to, there is no reason to suppose that the clone and the original individual will have the same psychology.

The insatiable hunger for power

This, it is feared, could lead to armies of clones being created, to act as enforcers for some Big Brother figure (Levin 1976).

The error is the same as before: it lies in thinking that cloning would lead to the "manufacture" of programmed, predictable human beings. This is effectively a misreading of Huxley's vision as depicted in *Brave New World* (Huxley 1932), and which is often cited in this context. The assumption about the "predictability" or "predetermined" nature of clones is generally valid where morphology is concerned, but erroneous in relation to psychology. The legions of alpha pluses and beta minuses or epsilons portrayed in the novel are not made up of morphologically identical individuals, but rather of individuals who are psychologically conditioned to their function and happy to perform it. These individuals are thus produced in the opposite way from clones, which are morphologically similar but psychologically different. They are, if you like, "anti-clones". As for the "armies of clones", should we fear their unleashing upon society, like so many futuristic storm-troopers, or are they rather to be seen as a symbol of hope, battling against the malevolent dictator in whose name they were created, and eventually destroying him?

One cannot help thinking either that, for all its dangers, the kind of biological cloning thus envisaged would probably cause less global, individual, social and human harm than much of the "ideological cloning" which occurred in the twentieth century, with all its attendant carnage.

78

Combination of ideological motivation – the quest for power and glory – and commercialism

This undoubtedly accounts for the behaviour of those various individuals who, in recent years, and amid much media fanfare, have announced their desire or determination to clone human beings: from Richard Seed in Chicago, of whom nothing more has been heard, to Severino Antinori in Rome and Panayiotis Zavos in Lexington (Kentucky), who are having problems finding somewhere to carry out their project. Other notable examples include the devoted followers of the former sports journalist Claude Vorilhon, who was inspired – seemingly by extraterrestrials – to found the Raelian sect. The Raelians, it must be said, are not to be lightly dismissed. Vorilhon himself is a man of some influence and his supporters include a scientist by the name of Dr Boisselier, various other specialists and, crucially, several dozen young women willing to donate their eggs, thus resolving of one of the main practical difficulties. While these women may be only too happy to bear their leader's clones, however, they may be less willing to repeat the favour for the sect's "clients". Furthermore, and for the same reasons as before, as well as the inevitable biological failures, there are bound to be psychosocial conflicts.

Commercialism and the popular imagination gone mad

There is an element of both in talk about the possibility of cloning Nobel prizewinners or great composers, thanks to some bizarre new kind of sperm bank of which, incidentally, one hears little mention any more. The same observation about the total unpredictability of clones' mental capacities is enough to relegate such talk to the realms of fantasy or charlatanism, of the kind once practised by peddlers of "moon dust".

The general feeling that emerges from examining these various manifestations of the popular imagination is that we should not allow ourselves to be intimidated by them, nor sidetracked into some futile dialectic discussion: as we have seen, the similarity between clones and original organisms is entirely illusory. The same applies to the supposed risk of undermining the diversity of the human species (quite safe in a world soon to have 8 billion human beings), the risk of inbreeding and, as we have also seen, the argument about the impermissibility of artificial predisposition. This last is already used in a "negative" sense to screen out diseases in prenatal diagnosis. More importantly, however, it is not concerned with the things that truly make us human, that is, our cerebral functioning, our intellectual capacities, our emotions, some might even say our soul.

There remains the question of the threat to human dignity, through the alleged instrumentalisation of the human embryo. Yet what is dignity? In the meantime, though, we will simply acknowledge that human dignity is priceless, that it is something which cannot be bought. But is not the same true of human life, the life of a loved one, a child for example? That is what we need to look at now.

Highly debatable scenarios that merit discussion

Dying spouse

The temptation, understandably, is to try to bring individuals back to life, to cheat death, as it were, by "recreating" them. Yet this is an insane and impossible dream, the implications of which need to be clearly thought through: scientists take a cell from a dying man and transfer its nucleus to the cytoplasm of one of his wife's enucleated eggs. They thus obtain an embryo which is successfully implanted in the uterus, the pregnancy proceeds as normal and the woman eventually gives birth to a child who will be her husband's clone. That is to say, he will be both her spouse and her child, without being fully one or the other. He will not be her child because apart from inhabiting her womb for nine months, he will have nothing to connect him to her, in particular none of her genetic material. Suppose, though, she nurtures and raises him as if he were an ordinary child. Is it not inevitable that, when he grows up, she will begin to see in him the very qualities that attracted her to her dead husband? How will she cope when he eventually marries? Even though he is not the same, even though he is twenty or thirty years younger and even though environmental factors will have left their mark, will she be able to stop herself trying to find in him the husband she lost? What possible good can come of such a situation, fraught with incest and frustration? To allow it to arise would be wholly indefensible.

Dying child

Some practitioners have argued that this is one instance where cloning could reasonably be regarded as a humane option. Ranged against them, however, are the legions of psychologists and sociologists who believe it would be unrealistic and indeed dangerous to try to create a "substitute" child, and who stress the importance of the grieving process.

Attempts along these lines have already been made, most notably by Dr Antinori who used a donated egg to enable a 60-year-old woman to have a baby after her 20-year-old son was killed in an accident. So far, the child shows no signs of any ill effects. This replacement child, however, was naturally very different from the one that died. What if he had been his clone? Would this be a mitigating or, on the contrary, an aggravating factor, particularly as the clone, although similar, would still be different, psychologically at any rate?

One possible justification for this ambiguous situation might be found in the highly specific case of a mother who, for one reason or another, was unable to conceive again, for example, because she had had her ovaries removed. Would it be justifiable to help such a woman become pregnant and give birth to a child resembling the deceased one, or would be it better to use, if possible, a donor egg or even to encourage her to forget about having a child altogether?

Whatever the case, one argument sometimes advanced, about the instrumentalisation of the child in order to satisfy some fanciful notion of motherhood, could hardly be regarded as tenable. Taken to its logical conclusion, such an argument would require us to take a long, sober look at the human urge to procreate in any circumstance, including, for example, in the case of a woman who conceives a child, naturally or artificially, in order to provide, say, a cord blood match for an older sibling suffering from leukaemia. The instrumentalisation and utilitarianism argument is therefore one that needs to be handled with care.

Artificial meiosis:
the artificial production of haploid cells (with only one set of chromosomes) from a diploid cell (containing a paired set of chromosomes). This process occurs naturally in the production of the germ cells.

Haploid cell:
a cell containing a single set of each chromosome (23 in humans). Germ cells are haploid.

Germ line:
any cell in the series of cells that eventually produce sperm and eggs.

Incurable male infertility

The problem here lies somewhere between the previous two. It is easy to see why a couple in this situation might be tempted to resort to cloning, particularly in an age when male infertility is becoming increasingly treatable, making it all the more distressing when the condition turns out to be incurable. The child would be the progeny of the couple, just as the clone of the deceased child was the progeny of its parents. It would presumably grow up in a propitious environment, even though it would bear no genetic relation to the mother, as in the case of an egg donation. It would, however, as in the dying-husband scenario, be a clone of the husband. It could, of course, be argued that the presence of the "progenitor", who would be more closely involved than, say, a husband who, in the same circumstances, had agreed to artificial insemination using a donor, more involved even than a real father, would help to ensure a certain balance within the couple. Set against this, however, are a number of potentially formidable problems: the ambiguous relationship between the mother and her husband's clone, the ambiguous feelings on the part of the progenitor towards this alter ego, who would be a living reminder of his own lost youth and the ambiguous feelings on the part of the adult clone towards his progenitor, a potentially devastating vision of himself in old age.

One wonders, however, whether it is altogether reasonable to focus attention on the use of cloning to treat infertility at a time when new medical options are beginning to emerge: not only *in vitro* maturation of germ cells, but also artificial meiosis* of the kind first mentioned years ago by G. David (Sureau 1999, 2000) and which involves constructing a "haploid" (23-chromosome) cell*, capable of fertilising or being fertilised, from a 46-chromosome, diploid cell, either a germ line cell* (Palermo 2001) or even a somatic cell. Perhaps one day artificial meiosis will provide the ultimate treatment for incurable infertility. It will certainly pose new, probably even more complex problems, including, for example, the problem of sex ratio, its potential imbalance and its demographic repercussions (Caselli and Vallin 2001), but it will also render the ideological debate about reproductive cloning obsolete.

For if there is one conclusion to be drawn here it is that the reproductive cloning at the centre of the above-mentioned concerns about the identity, uniqueness and diversity of clones, the motives of would-be applicants, be they individuals, couples or sects, and the potential effects on the genetic lottery and human dignity, should be treated with the utmost caution first and foremost because of the medical and psycho-social implications.

Mitochondria: energy producing bodies in the cell, outside the nucleus, which contain a small genome.

Over and above any ideological considerations, it is in these pragmatic terms that the issue of reproductive cloning should properly be viewed.

Some specific scenarios

Mitochondrial disease

This is caused by a genetic defect not of the nucleus, but rather of the mitochondria* (small bodies in the cytoplasm of most cells, which play a major role in cellular metabolism and activity). When the egg is fertilised, the disease has serious, often fatal consequences for the child and there is no known treatment.

The only solution that society can offer at present is egg donation, where the mother plays no part in determining her child's genetic heredity. An alternative route might to fertilise the carrier-woman's egg with her husband's sperm and then transfer the nucleus of one of the blastomeres of the resulting embryo into the enucleated egg of a woman who does not have the disorder, and implant this "neo-embryo" in the uterus of the carrier woman. There would be no cloning involved as the embryo transferred would be unique. Instead, what we would have would be a medically justified use of nuclear transfer technology.

Organ banks

The creation of clones by procreation for "organ banks" is a very common fantasy. However, it is, of course, morally indefensible and, what is more, unrealistic, because who would be the clone and of whose organs, therefore, would such a bank consist?

Transgenesis:
the introduction of new DNA sequences into the germ line, resulting in the production of transgenic animals.

Germ line gene therapy:
transfer of genes in the sex cells in order to genetically modify future generations.

Creating clones of embryos by somatic cell nuclear transfer raises, moreover, thorny scientific and ideological issues. Embryos are created by cloning, in order to obtain from their cells, isolated at the blastocyte stage, variously differentiated "lines" for use in human clinical practice, either for the cells themselves or in order to create new replacement organs. David (2001) observes, albeit tongue in cheek, that if mankind's dream of immortality had any chance at all of coming true, then it was more likely to be through this application of cloning than the kind of human reproductive cloning discussed elsewhere.

Pre-implantation diagnosis

This is often presented as a kind of biopsy used, in cases where there is a risk of passing on a genetic defect, to select healthy embryos and transfer them, the idea being to avoid prenatal diagnosis with its agonising wait and the trauma of termination. This is not the place to discuss the advantages and disadvantages of pre-implantation diagnosis. To call it a "biopsy", though, is something of a misnomer: a biopsy, after all, does not usually affect a quarter or half of the subject's body, which is effectively what happens here, since 1 to 2 out of 2 to 8 cells are analysed and hence destroyed.

It is important to recognise, indeed, that pre-implantation diagnosis is actually a form of cloning by blastomere separation, where one of the blastomeres (which, by definition, could have developed into an organism) is sacrificed in order to ascertain whether or not its artificial twin is healthy. This is basically a case of diagnostic horizontal cloning, now an accepted, legal practice. Which just goes to show that we should think twice before rushing to condemn.

Transgenesis* and germ line gene therapy*

It is a well-known fact that one of the key applications of animal cloning is transgenesis (Houdebine 2001). It is also a well-known fact that any modification of an individual's genetic heritage is prohibited under Article 13 of the Convention on Human Rights and Biomedicine. The French Parliament, moreover, faces the tricky task of bringing this provision into line with Article 16.4 of the French Civil Code (Law 653-94) that allows for the possibility of such modification when carrying out research into the prevention of hereditary diseases. It thus appears that, sooner or later, but probably later given the magnitude of the obstacles, governments will sanction genetic "manipulations" the purpose of which is to prevent not only the manifestation of a transmissible defect, but also the transmission itself, based on transgenetic animal experiments involving cloned embryo cells.

Would such "manipulations" fall within the scope of the Additional Protocol? At first glance, it would appear not, as the aim would be to correct a defect and thus create an individual whose genetic heritage would be different from that of the original; what is less certain, however, is whether it would be legitimate to carry out this kind of cloning if it involved attempting transgenesis on several cells at once.

What these various "scenarios" clearly show is that reproductive cloning is more a figment of the popular imagination than a genuine threat to humanity. It is quite possible that somewhere in the world, scientists have already produced such clones and if that is so, then it has to be said they have done less to disrupt the even tenor of our lives than the political, social, economic or even meteorological events that daily claim our attention.

That we should have the utmost reservations about such practices is, generally speaking, entirely proper. This is due largely to the medico-scientific risks, as revealed by animal experiments, and the potential individual, psychological and family repercussions, which simple clinical practice has served to highlight. Such considerations are infinitely more compelling than any ideology- or dogma-ridden theoretical speculation about the nature and the status of the embryo.

References

Caselli, G. and Vallin, J. (2001). "Une démographie sans limite". *Population*, p. 56. INED ed.

David, G. (2001). "Quel rêve derrière le clonage? reproduction ou immortalité." in *Juger la Vie*, 2001, Editions de La Découverte, Eds. Marcela Jacub et Pierre Jouannet, pp. 204-218.

Houdebine, L.M. (2001). *Transgenèse animale et clonage*. Dunod.

Huxley, A. (1932). (reprint 1955). *Brave New World*. Harmondsworth: Penguin Books in association with Chatto & Windus.

Levin, I. (1977). (New edition 1999). *The Boys from Brazil*. Penguin Press, London:

Palermo, G. 2.07.2001. *Livre des communications*. ESHRE (European Society of Human Reproduction and Embryology), Oxford.

Rorvik, D. (1978). *In his image: the cloning of a man*. Hamish Hamilton, London.

Salmon, C. (1998). *Des groupes sanguins aux empreintes génétiques*. France: Dominos, Flammarion.

Sureau, C. (1999). *Alice au pays des clones*. Stock. Ed. 2000. *Alice au pays des clones:* J'ai lu, Paris

Is human cloning inherently wrong?

by Professor Egbert Schroten

The birth of Dolly in 1996, the first mammal cloned by somatic cell nuclear transfer (see page 16), raised much public and, consequently, political concern. As soon as the Roslin Institute in Scotland had announced this biotechnological break-through,[1] Bill Clinton, the then President of the United States charged the National Bioethics Advisory Committee (NBAC) to report,[2] within only three months, on the consequences. And Jacques Santer, the then President of the European Commission requested the Group of Advisers on the Ethical Implications of Biotechnology (GAEIB) to write an Opinion on the ethical aspects of cloning techniques.[3] On a national level many governments acted likewise, and many churches voiced their concern as well, on ethical and theological grounds. It is interesting to note that public discussion immediately focused on human cloning. This is still the case, although the ethical and social aspects of animal cloning are sometimes discussed as well. The discussion on human cloning is stirred up by people like the Italian doctor Severino Antinori and the French biochemist Brigitte Boisselier, scientific director of Clonaid, who has already started attempts to make human clones. Throughout the world, however, there is a strong tendency to ban human cloning, because it is taken to be morally unacceptable. Laws and regulations have been and are being prepared and, recently, France and Germany took an initiative aimed at preparing a United Nations treaty to prohibit human cloning (see the chapter by André Albert, "Towards a world-wide ban on human cloning?").

Even if these and many other reactions may have a strong emotional side, it is worthwhile stressing that public concern should not be dismissed as being merely emotional. As far as ethics is concerned, emotional reactions constitute interesting material. What makes people concerned, angry, glad, proud and the like? It is because they think that there are values at stake. Emotions are "markers of values" and the task of ethics is to analyse them and to put these values on such a level that we are able to discuss them. In this respect an ethicist can be

1.
http://www.roslin.ac.uk/

2.
Although the Charter of the NBAC expired on 3 October 2001, information on its reports, findings and meetings can be accessed at: http://bioethics.georgetown.edu/nbac/

3.
http://europa.eu.int/comm/european_group_ethics/gaieb/fr/biotec02.htm and http://europa.eu.int/comm/european_group_ethics/gee1_fr.htm

Monovular twins:
twins produced from
the splitting of a
single egg. Also
known as mono-
zygotic twins.

1.
I shall use here two
Opinions of the
Group of Advisers on
the Ethical Implica-
tions of Biotechnology
to the European
Commission (GAEIB),
which later became
the European Group
on Ethics in Science
and New Technolo-
gies to the European
Commission (EGE). I
was a member of
both of these groups
until recently, namely
the group which pre-
pared Opinion No. 9
on Ethical Aspects of
Cloning Techniques
(May 1997) http://
europa.eu.int/comm/
european_group_
ethics/gaieb/en/
opinion9.pdf and the
group which drew up
Opinion No 15 on
*Ethical Aspects on
Human Stem Cell
Research and Use*
(November 2000)
http://europa.eu.int/
comm/european_
group_ethics/docs/
avis15_en.pdf.

2.
To avoid misunder-
standing: This does
not mean that animal
cloning is not morally
problematic. On the
contrary! See, for
instance GAEIB
Opinion No. 9, para-
graphs 2.1-2.3.

compared to a psychoanalyist, who digs up values, principles
and norms which are, more or less subconsciously, present in a
person, a community, a society, a culture, and tries to put them
on a level where they can be discussed in a rational way.

So, cloning is apparently taken to be morally problematic. In
this chapter I want to deal with the question of why this is the
case and to analyse some of the main arguments. But prior to
this it is important to clearly define what we are talking about.
Therefore I want to make some preliminary remarks on
cloning and on ethics.[1] Cloning is the technology of producing
genetically "identical" organisms. I have put quotation marks
around the word "identical" because organisms are never
completely identical, as can be seen in the case of monovular
twins*. To avoid any misunderstanding, it would therefore be
better to avoid the word "identical" in this context and to speak
in terms of "similarity". Cloning may involve embryo splitting*,
or it may involve the transfer of a (somatic) cell nucleus, in
which case the nuclear genes are "identical" whereas a small
number of genes which are outside the nucleus in the cell (the
mitochondrial genes*) are not. I shall not discuss further here
the technology of cloning, because this will be done elsewhere
in this book. Suffice it to say that this technology is still at an
experimental stage, which, incidentally, gives us the opportunity to take time in reflecting on this issue.

I want to concentrate on the issue of cloning by means of
somatic cell nuclear transfer (SCNT) in the context of human
reproduction.[2] It is important to draw a clear distinction
between reproductive cloning which aims to create (similar)
human individuals, and so-called "therapeutic cloning" in
which nuclear transfer technology is used to obtain (human)
embryonic stem cells for therapeutic purposes. Stem cells are
cells that are not (yet) fully differentiated and therefore can
reconstitute one or several types of tissues. The challenge is to
be able to control the differentiation of these human stem
cells. If this kind of research becomes feasible, it will not only
be of great interest for basic developmental biology and any
spin-off, but also for therapeutic transplantation*, for instance
in the case of heart diseases, diabetes, hepatitis, Parkinson's
disease, burns, and some forms of cancer. It is therefore not

surprising that there is, within the scientific community, great pressure to intensify this kind of promising research.

However, I want to focus on the first kind of cloning, namely SCNT in the context of human reproduction, which means that I shall deal with therapeutic cloning only when there is an overlap from an ethical point of view. But before doing that, we have to establish what we mean by ethics. Ethics is about systematic moral reflection. More specifically, it is about reflecting on what we ought and ought not do and why this is the case. Such reflections are carried out systematically, that is, methodologically, in the light of moral theory, a more or less consistent framework of principles, norms, values and virtues. However, let there be no misunderstanding. Ethics, in particular applied ethics (for instance bioethics), is not only about norms but also about facts and concepts. If we are in a problem situation, for instance human cloning, and the question is what ought (not) to be done, we have to clarify what we mean by the concept of human cloning as well as obtain some information about what is "state of the art" in this area of biomedical technology, that is, what is going on and the possible consequences of this for society. Clarification and more information on this issue is essential if we are to see which moral principles, norms and values are relevant and in what ways they are relevant.

Conceptual clarification can be seen as a typical philosophical exercise, but analysis of a problem situation is not, necessarily. Since factual information is vital for an ethicist, it is clear that applied ethics (like medical ethics or bioethics) is a multidisciplinary undertaking. For factual information an ethicist needs the help of experts in other disciplines (for example, biology and biomedicine). This is not always as easy as it seems because there is, in practice, no such thing as "bare facts". Interpretation of factual information plays an important role in the communication between science and ethics and even if both agree on the interpretation of a state of affairs, there may be a difference in moral evaluation, for we live in a pluralist society which means, to say the least, that there is not just one framework of principles, norms, values and virtues. In our so-called post-modern society, people choose to lead different ways of

Embryo splitting: division of a single embryo into twins or quadruplets.

Mitochondrial genes: energy-producing bodies in the cell, which contain a small genome.

Therapeutic transplantation: the transfer of tissues, which have been grown from stem cells *in vitro*, back into the body for therapeutic purposes.

life and they justify their choices (if at all) through a variety of different arguments.

However, without denying the problems of pluralism, these problems should not be exaggerated, certainly not in Europe. European culture is based on the Judaeo-Christian and Humanist tradition. I do not state that this is an exclusive way, but rather I am using it as a basis to indicate where we need to look for the roots of our cultural tradition. As far as I can see, there is a significant overlap in Humanist/Christian and Muslim values, which should be taken into account to a greater extent by public policy. This is why I want to underline that pluralism is not so much a question of having totally different values, but rather a question of giving different priorities to values. Concerning euthanasia, for instance, the Pro-life movement does not deny the importance of human freedom and responsibility, and the Pro-choice movement does not deny the importance of the protection of human life, but the two groups disagree as to which values should have priority. This holds true for many moral dilemmas in biotechnology. There is, at least in our culture, a fundamental sharing of basic principles, norms, values and virtues which creates a common ground for an ethical discussion. That is why, for instance, Article 7 of the EU Fifth Framework Programme for Research, Technology and Development,[1] technological development and demonstration activities (1998-2002), requesting compliance with fundamental ethical principles, is perhaps vague but not nonsensical. For me, this is the reason why it is relevant to refer sometimes to (arguments taken from) western cultural and religious tradition.

Ethical objections

After making these introductory remarks about cloning and ethics, it seems an appropriate time to pose the following question: "Why is human cloning taken to be morally problematic?" I want to take as my starting point Opinion No. 9 of the GAEIB, published in May 1997. Concerning human reproductive cloning the Group writes:

> "2.6. As far as reproductive cloning is concerned, many motives have been proposed, from the frankly selfish (the

1.
http://europa.eu.int/
comm/research/fp5.
html

elderly millionaire vainly seeking immortality) to the apparently acceptable (the couple seeking a replacement for a dead child, or a fully compatible donor for a dying child, or the attempt to perpetuate some extraordinary artistic or intellectual talent). Considerations of instrumentalisation and eugenics render such acts ethically unacceptable. In addition, since these techniques entail increased potential risks, safety considerations constitute another ethical objection. In the light of these considerations, any attempt to produce a genetically identical human individual by nuclear substitution from a human adult or child cell ('reproductive cloning') should be prohibited.

2.7. The ethical objections against cloning also rule out any attempt to make genetically identical embryos for clinical use in assisted reproduction, either by embryo splitting or by nuclear transfer from an existing embryo, however understandable.

2.8 Multiple cloning is *a fortiori* unacceptable. (....)

2.9 Taking into account the serious controversies surrounding human embryo research: for those countries in which non-therapeutic research on human embryos is allowed under strict licence, a research project involving nuclear substitution should have the objective either to throw light on the cause of human disease or to contribute to the alleviation of suffering, and should not include replacement of the manipulated embryo in a uterus.

2.10. The European Community should clearly express its condemnation of human reproductive cloning and should take this into account in the relevant texts and regulations in preparation ....".

What we see in this Opinion is a clear condemnation of human reproductive cloning and advice to the European Community on how to express this condemnation in its policy measures. Cloning in the context of non-therapeutic research is conditionally allowed in certain countries but only under strict regulation. This position has a peculiar implication: it implies that the use of the cloning technology in non-therapeutic research is allowed in those countries which permit it and is not allowed in those countries which prohibit it. This may be a wise political statement, but from an ethical perspective it is not very

IVF
(*in vitro* fertilisation): biological process that occurs in a container outside of the organism e.g. cells or tissues grown in culture.

satisfactory. A second remark is that the arguments against human cloning may be divided roughly in two kinds:

- practical (or utilitarian), looking at ends and consequences;
- fundamental (or deontological), looking at norms and principles.

The practical arguments could be summarised by the fact that human cloning is too risky, because the technology is in its infancy. I fully agree with this as it is important to be realistic. Dolly was born only after almost 300 attempts. The technology could, of course, be improved, but still we will need too many human embryos and our knowledge about the implications for the developing embryo and the child to be born is very limited. The same holds true for our knowledge of the social, ethical and religious aspects. In line with the classic rule *in dubiis abstine* (withdraw in cases of doubt) and for safety reasons, many people, in the community of science as well, are requesting that there be a moratorium on human reproductive cloning. Much more research should be done before we can think of moving into human cloning by SNCT.

But suppose cloning does become feasible and safe, not absolutely safe of course, for that is impossible in practice, but, let us say, as safe as IVF* in general. Then we are still confronted by the fundamental arguments. In the Opinion quoted above, something is said about "apparently acceptable" and "unacceptable" motives and about "considerations of intrumentalisation". In addition to what has been said in the Opinion, one may, as to the motives, point to the fact that in science and technology the quest for knowledge of how and why something works and the challenge of discovering and mapping new research areas is always present. In view of human cloning one may come across motives;[1] like:

- infertility (and therefore the desire for offspring);
- the wish to have a child to replace another child that has died; and
- the wish to have a child to provide organ or tissue for a relation as this would dramatically minimise the risk of rejection;

1.
See Theo Boer, "Het kloneren van mensen. Een theologisch-ethisch commentaar" ("Human cloning. Comments from the perspective of theological ethics") in Dick G.A. Koelega & Willem B. Drees, *God & co?. Geloven in een technologische cultuur (God & Company? Faith in a technological culture)*, Kampen/Driebergen (Kok/MCKS) (2000), pp. 179-199.

- a desire to have a clone of oneself (this could be very disappointing because the clone will certainly not be the same person);
- the extreme desire to have multiple cloning (with an eye on *The boys from Brazil*[1]); and
- a wish to enhance the quality of the offspring, for instance, with regard to their health, intelligence, or beauty (eugenics).

Looking around in the relevant literature, one could say that one of the fundamental arguments against human cloning is that it is often seen as a violation of "natural order" in general and of human dignity in particular. In this context the moral status of the human embryo is used as an argument against human cloning. These arguments aim at the total rejection of human cloning, because it is taken to be intrinsically wrong. I want to further develop these fundamental arguments against human reproductive cloning*, because if we have to say "no" in principle, we have to say "no" in practice as well.

A violation of human dignity ?

The question is then: "Does human reproductive cloning constitute a violation of human dignity?" The first thing to say is that "human dignity" is a rather vague concept. Could we be more precise concerning its meaning? In order to answer this question, I will refer to Gilbert Hottois' analysis in the *Nouvelle encyclopédie de bioéthique*[2]. With reference to the philosopher Kant he mentions three important characteristics of human dignity, in close connection to each other, namely: autonomy; identity or singularity; and subjectivity or freedom. Focusing on two motives mentioned above, namely infertility and the wish to replace a child that has died, his critical question is whether reproductive cloning necessarily implies the negation of the singularity (or identity), the autonomy, and the subjectivity (or freedom) of the clone.

As to the identity or singularity issue, his argument starts by underlining a point that has already been made above, namely that the clone(s) and its (their) source are not biologically identical. Due to the mitochondrial DNA and the interactions between the genes, and between the genes and their environment,

Human reproductive cloning: production of a human being that is genetically identical to another (by nuclear substitution from a human adult somatic cell or child cell, or by artificial embryo splitting).

1.
The boys from Brazil (1978) Science fiction novel by Ira Levin Random House; New York.

2.
Gilbert Hottois and Jean-Noel Missa (eds), with the collaboration of Marie-Geneviève Pinsart and Pascal Chabot, *Nouvelle encyclopédie de bioéthique*. Brussels (De Boeck Université) (2001), s.v. "Clonage humain réproductif"

during the development of the clone, which could carry muta-
tions, there will always be differences, as we can see in the case
of monovular twins. Therefore I have already suggested that we
speak of "similarity" rather than of being "identical". Moreover,
this similarity, however great it may be, can be seen only as part
of the individual's "identity", because a human's individuality is
shaped by psychological, social and cultural factors as well. In
the case of cloning, this aspect will even be increased by the difference in age between the clone and its source. According to
Hottois the problem of individual identity depends largely on
how the clone is accepted by his/her family and society and the
auto-perception of the clone as a consequence thereof.

Does reproductive cloning imply a negation of human autonomy? Here, Hottois looks at the fear of clones being used for
instrumental purposes and agrees that one could imagine
cases where this fear would be justified, namely when clones
would be used as mere reservoirs of organs or as material for a
eugenics project. But he underlines that he is still focusing on
the above-mentioned motives for reproductive cloning. He
points out that these cases do not differ very much from normal situations, where parents have ideas about (the future of)
their children, have influences on them by bringing them up,
by schooling and so on. We all know that there are excesses in
these situations as well, for instance indoctrination. In short,
the instrumental aspect is always present to a certain extent in
human relations, even in friendship and love. This does not
hamper human beings from becoming autonomous subjects.
Evil is not in the technology itself but in the way we use it and
in the way we treat our fellow creatures. Kant does not deny
instrumental aspects in the human situation, but he says in his
famous maxim that we have to relate to other human beings
not merely as means but always also as ends. There is no reason to think that reproductive cloning would necessarily imply
the impossibility of clones becoming autonomous subjects.

But is there not still a violation of human freedom and subjectivity because of the elimination of the genetic "lottery" in
reproduction with its inexhaustible uncertainty, which can be
seen as a protection of the human being? Hottois points out
that there are some untenable presuppositions in this argument.

In the first place there is the suggestion that cloning, in contrast with sexual reproduction, implies a form of determinism. He rejects this suggestion with reference to the arguments he used in the context of the issue of identity or singularity. These arguments also hold true with regard to the idea that a clone would be, in his/her personality, an exact copy of the personality of the source. He stresses that the development of a clone, as a free human subject, depends largely on the contexts of the family and the society in which (s)he lives. He also rejects the presupposition that human freedom depends on indeterminism, in this case the genetic lottery, which would imply that any intervention in this lottery would constitute a violation of human freedom. In the western philosophical tradition, there is a vivid discussion on the issue of freedom and (in-)determinism, and one could (and perhaps should) say more about it than Hottois does, for instance by taking into account the significance of human corporality (finitude) for human freedom. But he is right in saying that human freedom and reason are closely linked and that it is not correct (Hottois uses words such as "absurd" and "contradictory") to suggest that human freedom is dependant on indeterminism.

Hottois underlines that his analysis should not be seen as a vindication of cloning. He is arguing against the objection that human reproductive cloning constitutes a violation of human dignity – a point with which I am in agreement. One can imagine cases where cloning does constitute a violation of human dignity, but it does not always have to be. What Hottois, in his contribution, does not take into account, however, is the issue of the moral status of the human embryo, which is often discussed in this context as well. The GAEIB, in the Opinion quoted above, rightly points to the fact that in cloning technology research on human embryos is presupposed. In the following section I will look at this further. Because this issue plays an important role in therapeutic cloning as well, it would be inaccurate to neglect this aspect of the problem.

The human embryo

One of the main traditional philosophical and theological positions held in the western world today concerning the embryo,

Cryopreservation:
preservation of tissues or organisms by freezing at very low temperatures.

may be summed up using the words of Tertullian, a philosopher/theologian of the second century, who wrote: *Homo est et qui est futurus*, a future human being is a human being as well. This means that in practice an embryo has to be treated as a human being, in other words as if it had already been born. There is a lot more to be said about Tertullian and the Judaeo-Christian mainstream tradition, which has greatly influenced western philosophy. However, Tertullian does not always seem to be consistent in his view on the human embryo and there is, in the Judaeo-Christian tradition, much discussion on the issue of the soul. This discussion takes place in the context of the problem of abortion and against the background of the merging of Aristotelian embryology and theological theories concerning the creation of the human soul. The combination of these two lines of thought resulted in an ongoing discussion as to when the human embryo/foetus can be seen to have a soul – a vital moment in becoming a human being. However, I do not want to elaborate on these issues in the history of ideas here, but I would like to use Tertullian's saying to represent the classical mainstream philosophical and theological position, not least because it has been taken on board by the Vatican and the Pro-life movement.

So *homo est et qui est futurus*, a future human being is to be treated as a human being as well. What can be said about this position? Very roughly, my answer would be as follows: in the western cultural tradition, the issue of the moral status of the human embryo has been discussed in the context of the problem of abortion and of obstetrics.[1] The embryo meant the embryo in the womb, a future child *in spe*. However, since the technologies of IVF and, later on, cryopreservation* have become realities, we have to face a new situation or, rather, new facts: there are embryos *in vitro* and *in the freezer*, which means that they are not in the womb. IVF then, as with other technologies like cloning, presents us with an extension of our responsibilities; this means, in practice, an extension of the realm of decision-making. As long as these embryos are intended to be transferred to a uterus, as will be the case in reproductive cloning, we could see them, and for that matter treat them, as future children *in spe*. But, if for one reason or

1.
See J.G. te Lindert, *Over de status van het menselijk embryo in de joodse en de christelijke ethiek* (About the status of the human embryo in Jewish and Christian Ethics). Utrecht (PhD thesis) 1998.

another, we decide not to transfer them into a uterus, as is the case in therapeutic cloning, they are still human embryos but not future children *in spe*. Thus, IVF and cryopreservation make it necessary to distinguish between two categories of human embryos, according to what they are to be used for: embryos to be transferred and embryos not to be transferred (so-called "surplus" or "supernumerary" embryos*).

Supernumary embryos: surplus embryos, resulting from IVF treatment, which are not implanted in the womb.

Nidation: implantation.

As surplus embryos are not destined to become children, their moral status differs from embryos that are destined to become children. Although they share human genetic inheritance, which implies that they have a special status, they will not become human beings in the normal sense of the word. This means that, although they are in a special position, surplus embryos need not be treated as future human beings. Their moral status is lower, in my opinion, than that of embryos which will later become children.

A counter-argument often used against this position is what I would call the "ontological status" argument. The attempt to make a distinction based on the moral status of the human embryo, that is, relying on the intended use of the embryo argument, does not take into account the ontological status of the embryo, that is, what an embryo really is. Since all embryos share human genetic heritage, they are, after the process of fertilisation, potential human beings. It would be misleading of me to refute this. This is why I think it is important for me to acknowledge that surplus embryos should hold a special status. However, I do not think that it is a strong argument against my position that it is what embryos are to be used for which is morally relevant; it is this which should be taken into account in policy-making. Let me explain: the potential of the human embryo to become a human being is certainly there, but it is very unlikely, if I may say so, because it depends so heavily on external factors, in other words on conditions which have nothing to do with the status (the "ontological status") of the embryo itself. These conditions include:

- successful transfer into a womb;
- nidation*;
- and a healthy pregnancy.

If these conditions are not met, an embryo will never become a human being in the normal sense of the word. Quite the contrary. If a (human) embryo is left alone, it will die. In other words, sharing human genetic heritage is a necessary but not a sufficient condition for becoming a human being. And since supernumerary embryos do not meet these conditions they will not become actual human beings and should not, therefore, be treated as such.

Or should they? In the wake of the ontological status argument there is also the right to life argument. Every (any?) human embryo (potential human being) has the right to life (to become an actual human being), which means that they should be exposed to the conditions mentioned above (transfer into the womb, nidation and pregnancy). Let us have a closer look at this argument, which I do not find particularly convincing. And let us, for the sake of argument, avoid the difficulties of the "language of rights" in the context of embryos (fertilised eggs). The argument is, then, the (moral) claim that any human embryo should always have a fair chance of becoming a human being. The question is: "Why?" And perhaps also: "How?" This is an open moral question. For, if we argue from the position of the ontological status of the human embryo, we would not have a very strong case, as we have seen above. And if it is indeed an open moral question, it would not be difficult to launch a *reductio ad absurdum* argument here, for instance in view of the fact that, in nature, probably more than 60% of fertilised eggs are lost, or in view of the fact that there are presumably hundreds of thousands of supernumerary embryos in Europe.

Perhaps one should mention, in this context, a religious argument as well: The human embryo is God's creation. I shall limit myself to two remarks here.

1. The theological question is: "What does that mean in practice"? Creation is a difficult concept to bring into operation in the discipline of ethics, partly because this concept is not clear, partly because it is applied in various ways, at least in Christian tradition. If it is taken to mean that we should respect human embryos, it is not of great help here.

2. The next question would be: "How? In what way?" However, in the Christian tradition (and, as far as I know, in the Jewish and Muslim traditions), special attention is paid to the creation of the soul. Mainstream thought in these religions is that the (rational) soul is created forty days after conception. Aristotle's influence here is unmistakable, but I shall not elaborate on this. What I want to say here is not that we should go back again to an embryology of the old days, but that mainstream western religious tradition purports that the soul is not created immediately after conception, implying, as far as I can see, that human dignity is not to be attributed immediately after conception.

To culture embryos: to grow embryos outside of the body *(in vitro)*.

As to the moral status of the human embryo, I have focused on supernumerary embryos. But we are all aware of another issue in this context, namely the issue of culturing embryos* specifically for research. Although here, from the beginning, the intention is to use human embryos for research and not to transfer them to the womb, making this procedure even more morally problematic, I would nevertheless claim that my position holds true in this context as well. These cultured embryos, too, will belong to the same category as supernumerary embryos. This would mean, for instance, that although one could speak in this context of an "instrumentalisation" of human life, it is not a question of using human beings as instruments in the usual sense of the word.

In short, what I want to say is that the distinction between human embryos on the basis of their intended use, is morally relevant. This claim is, as far as I can see, not refuted by arguments based on the ontological status of the embryo, or on the basis of an embryo's right to life, or on the basis of mainstream Christian thought. So, being a Christian theologian, I want to underline that, as far as I can see, my claim is in accordance with principles of upright Christian ethics. Science and technology (*in casu* IVF and cryopreservation), whether we like it or not, have led us into new situations, which force us to rethink traditional moral principles and values. Of course, we should be critical but we should not bury our heads in the sand and refuse to face facts.

My conclusion is that, from a moral point of view, supernumerary embryos, designed for research purposes only, are to be differentiated from embryos which are intended to be transferred to a uterus. This distinction could become the basis of any public policy concerning the use of human embryos for research. However, as these embryos do share human genetic heritage they should be considered to belong to a special category, and only under strict conditions should they be used for research. These conditions would include, for instance, there being no alternatives and the research aims being substantial and morally acceptable. These and other conditions would underline the special status of the human surplus embryo and in these terms the expression "adequate protection", as it is used in the Council of Europe's Convention on Human Rights and Biomedicine,[1] should be implemented.

Unnaturalness

One fundamental argument, however, is still left, namely the argument that cloning is unnatural. It is clearly a high-tech issue, and thus highly artificial. The question is, however: "Does that matter?" Does nature teach us how to behave? Does evolution or (belief in) creationism imply a moral order in nature? Does cloning mean that we are violating the fundamentals of this "given" order of nature?

This "natural law" stance holds a strong position in western culture. It is not the place here to reopen the very sophisticated philosophical and theological discussion surrounding the issue, although biomedical technology may give plenty of reasons to do that. Apart from the fact that technology is by definition more or less artificial, and thus "unnatural", the point I want to make here is that natural law is not an unambiguous guide. On one hand, to use the Christian version of it, if we believe that God created the world, the implication is that this world is somehow an expression of His will. On the other hand, however, because we are confronted with much evil in nature, there is always a large element of human construction and interpretation in the way we "perceive" moral order in nature. In other words, natural law is mainly human interpretation in

1.
Full title : The Convention for the Protection of Human Rights and Dignity of the Human Being with regard to the Application of Biology and Medicine. Article 18.1 reads : "Where the law allows research on embryos *in vitro*, it shall ensure adequate protection of the embryo.".

the light of (increasing) knowledge and faith. Therefore it is, to a large degree, "cultural" law.

In the light of this, I would also point to something else. Perhaps it is an exaggeration but my impression is that there is some ideology emerging, which, perhaps implicitly, evaluates nature as intrinsically good, at least much better than human intervention. Nature is seen as having a sort of taboo status, something which is nearly divine, and human intervention becomes almost, by definition, an infringement of the "sanctity" of nature. If this is the case, than we are confronted with a form of "neo-paganism". The concept that nature is divine is rejected in western culture, at least in the Judaeo-Christian tradition. The world is looked upon as God's creation, which means that it is not-God. Moreover, this world is spoiled by sin and evil. And if we take another interpretation of creation, which is present in the Bible, namely that creation is not *creatio ex nihilo* (creation out of nothing) but God creating cosmos out of chaos, which means that chaos is a constant threat to and in cosmos, then the idea of moral order in creation is even more difficult to maintain. In short, as may be clear by now, in my view the natural law argument against biomedical technology is not a very strong one.

Is human reproductive cloning intrinsically wrong? Analysis of the fundamental arguments which are put forward in opposition to this technology do not necessarily force us to make this conclusion. It may be morally problematic, but if it would be safe it does not seem to be a priori morally unacceptable. In other words, the technology is not "in itself" wrong, it depends on what it is used for. If it can be made safe and if it were used in the context of infertility treatment or prevention of serious diseases, it could be placed on the same level as advanced IVF technology, like ICSI (IntraCytoplasmatic Sperm Injection).

However, the technology is not (yet) safe. Therefore I would support a moratorium on human reproductive cloning. But because human cloning is not intrinsically wrong, I personally cannot see why research in this area should be prohibited. It could be allowed under strict conditions, because cloning is morally problematic and it could be used for unacceptable ends, for instance eugenics. In line with the GAEIB and the EGE, I would therefore plead for a policy of great prudence, which I would translate here into a so-called "no unless policy", which means that human cloning is forbidden unless there are convincing arguments to lift this ban. Such a decision should be taken under the supervision of an independent multidisciplinary licensing body, which would also represent public concern. Developing a moral framework would be the first thing to do.

Finally, there is of course the question whether, in society, we should move at all in the direction of reproductive cloning. This is, ultimately, a political question, related to the allocation of financial means, the question of the price (literally and metaphorically) we want to pay for the treatment of infertility and some diseases, the place of science and technology in the context of human reproduction, as well as in society as a whole. These are fascinating issues with many (other) ethical, philosophical and religious aspects. I do not want to elaborate on them further here. Let me make three final remarks to put this technology into perspective: the first is that, however sad it certainly may be not to have children when you would like them, human life can be fruitful without having children. The second is that science and technology, in the end, are not there for their own sake but to contribute to "the good", which means that they should be carried out in compliance with ethical principles in society. The third is that ethics is not an eternal, static, monolithic system, but a discipline which is rooted in tradition but at the same time in open and dynamic interaction with what is going on in society, and therefore in science and technology, as well.

"Therapeutic" cloning and the status of the embryo

by Professor Axel Kahn

For hundreds of years, the medical world has dreamed of being able to repair, piece by piece, any ailing or worn-out part of the human body. This dream began to come true with the first organ transplants carried out some fifty years ago. Since then, improved surgical techniques and the development of new immuno-suppressors* have led to an increase in the number of situations where transplants are feasible and have boosted transplant success rates. Today, the main problem is the severe shortage of available grafts and this explains the current interest in the use of organs from animals (xenotransplantation) once compatibility problems have been addressed by transferring human genes to the animal. However, there is still a long way to go before the difficulties of immune rejection can be overcome. This has led to growing interest in cell therapy which has its own specific applications and which could in some cases be an alternative to organ transplants.

In the future, *in vitro* tissue engineering techniques could in all probability even produce certain organs in culture which could then be transplanted; this is already happening on an experimental basis with the skin, blood vessels, the bladder and the cornea. Cell therapy is today widely used for blood disorders (transplantation of blood-forming stem cells) and burns (skin grafts), and experiments are currently being carried out involving the transplants of liver cells to treat liver diseases, endocrine pancreas cells to treat diabetes and neural stem cells to treat various neuro-degenerative disorders. With the exception of skin grafts and medullary autografts*, the techniques used have involved the transfer of allogenic cells, that is, cells from a donor who is not the recipient. More often than not, the cells are of foetal origin. There is still of course the problem of immune rejection, which is at its lowest in transplants in the central nervous system because of its unique immunological status. The availability of foetal cells is closely linked to termination of pregnancies, which obviously raises ethical concerns.

Immuno-suppressors: substance which contributes to the partial or complete suppression of the immune response of the individual.

Medullary autografts: transplantation of bone marrow within the same patient.

Multipotent:
capable of developing into several different cell types.

Blastocyst:
the hollow sphere of cells that develops from the morula (the solid mass of cells produced by the first cell divisions of a fertilised egg) and implants in the uterine wall.

Differentiate:
develop into the final specialised structure and function of the cell.

Histocompatibility:
the ability of a tissue to be accepted or rejected if transplanted into another individual due to certain proteins present on the surface.

In any event, cells from this source are unlikely to be in sufficient quantity to satisfy the possible requirements of cell therapy in the future.

Regenerative medicine and stem cells

In 1998 a new technique was developed based on the culture of mice embryonic stem cells, perfected ten years earlier. Human multipotent* cells were successfully cultured using human blastocysts* or primordial germ cells (Thomson *et al.* 1998; Shamblott *et al.* 1998). Clearly, such cell lines could solve the problem of the low availability of foetal tissue. However, the problem of immune rejection remains; furthermore, there is still a long way to go before it will be possible – in mice, and even more so in humans – to influence at will these embryonic stem cells to develop ("differentiate"*) into the type of cells needed for transplantation. Numerous checks also have to be made to ensure that the cells in question are not likely to form tumours.

The problem of immune rejection could of course be solved if these embryonic stem cells came from embryos cloned by transferring the nucleus of a cell from the patient into a receiver oocyte (a female germ cell) from which the nucleus had been removed (Solter 1999). Without losing sight of the ethical concerns surrounding the creation by cloning of human embryos for the purpose of cell therapy, discussed below, it might also be possible to "personalise" human embryo stem cell lines by replacing their nucleus with one taken from the patient's own cells. It is also perfectly feasible to create banks of human embryonic stem cells produced from tens of thousands of surplus embryos, frozen after *in vitro* fertilisation, which were not to be used for reproductive purposes and which would otherwise be destroyed. These colonies would be carefully classified according to the major histocompatibility* complex antigens, as is currently the case for collections of cord-blood stem cells. In addition, genetic manipulation could help improve immunological tolerance to these cells.

Lastly, recent research has shown that it might not be necessary to rely on embryonic cells, since cells which are able to regenerate are to be found in all adults. Furthermore, these stem cells, originally thought to be unable to change course and produce other types of cells, are in fact much more versatile than first imagined. Numerous articles have been published showing that there are neural stem cells in the sub-ventricular zone* of the central nervous system in adults which continue to create new neurons* (Doetsch *et al.* 1999; Goudl *et al.* 2000). In some experiments, these neural stem cells have been able to differentiate into various types of cell (Clarke *et al.* 2000). Blood-forming stem cells are able to generate not only the different types of blood cells but also muscle (Ferrari *et al.* 1998), liver (Petersen *et al.* 1999) and even nerve cells. It is possible to isolate cells in the muscle which are able to repopulate the bone marrow (Gussoni *et al.* 1999). Finally, bone marrow also contains mesenchymal stem cells* which can generate cartilage, bone, tendon, adipose tissue, etc. (Pittenger *et al.* 1999). The solution to cell therapy may therefore lie in each and every one of us.

At present the capacity of these specific stem cells to regenerate the tissue from which they derive and particularly to differentiate into other types of cells, is clearly insufficient, probably because the technique has not yet been perfected. This would require the use of growth factors (cytokines) and survival factors to ensure that each cell population could proliferate and differentiate into the desired cells. It is highly probable that the extremely rapid development of functional genomic programmes* following on from the sequencing of the human genome will increase our knowledge of the full range of factors enabling us to control tissue regeneration from minority populations of stem cells. In the future, therefore, there is a high chance that the combination of these expansion* and differentiation factors, either directly *in vitro* or *ex vivo*, applied to cells in culture will considerably increase the possibilities for regenerative medicine. In certain cases, the effectiveness of cell repopulation will require transgenes* to be added to the cells thus stimulated in order to give them a selective advantage. The viability of these approaches has been shown for the regeneration of blood tissue (Sauter *et al.* 1997; Bunting *et al.* 1998) and the liver (Mignon *et al.* 1998). Clearly,

Sub-ventricular zone:
the area around the ventricle of the brain which continues to produce new nerve cells after birth.

Neuron:
cells which can create nerve cells.

Mesenchymal stem cells:
undifferentiated embryonic cells from tissue (the mesoderm), that gives rise to muscle, connective tissue, blood, etc. which can differentiate into several different tissue types.

Functional genomic programmes:
research programmes aiming to assign a function to all genes in the human genome.

Expansion:
the ability of stem cells to develop into several different cell types.

Transgenes:
foreign genes that are introduced into an organism by injecting the genes into newly fertilised eggs. Some of the animals that develop from the injected eggs (transgenic animals) will carry the foreign genes in their genomes and will transmit them to their progeny.

this approach could be used in conjunction with other types of gene transfer into cells in culture, either to improve the quality of regenerating cells or to assign to them a therapeutic role that physiologically they do not have.

Given the range of and linkage between these techniques, it is perfectly reasonable to anticipate a genuine revolution in treatment, which – as a result of a dramatic extension in the number of situations where cell therapy can be applied – will compensate in a growing number of cases for age or illness-related malfunctions of virtually all organs. However, above and beyond the theory that still has to be confirmed, some of these approaches raise significant ethical questions.

Therapeutic cloning

Cloning human embryos could have two aims, the first to provide treatment and the second for the purposes of reproduction. In the first case, embryonic cells, which in genetic and immunological terms are identical to those of patients waiting for cell transplants, need to be obtained in order to treat a large variety of diseases: neuro-degenerative disorders such as Parkinson's disease or Alzheimer's disease, cancers, diabetes, liver failure, burns, etc.

Carrying out such a programme would require biologists to be able to differentiate the isolated stem cells of a cloned embryo that, as we have seen, is far beyond the current state of the art, but is not impossible. In the future, a person suffering from Parkinson's disease or diabetes could ask his wife or daughter to donate oocytes or obtain them from paid donors. The doctor would replace the nucleus of these ovules with a nucleus from any cell of the person concerned and would cultivate the cloned embryo for several days under laboratory conditions. After about day 7 the embryo consists of a hollow sphere of a hundred or so cells, the blastocyst, inside which there is a small cluster of cells which will develop into the foetus strictly speaking. These cells can be isolated and cultivated. They retain the ability to differentiate into all the various types of cells to be found in the adult; they are called embryonic stem cells or ES cells. If the doctor is able to influence them, by manipulating the culture conditions, to differentiate into brain cells or pancreas cells, they can then be transplanted into the patient in order to treat the Parkinson's disease or diabetes. There should be no problem of rejection because the cells introduced will be identical to those of the person receiving them.

It must be pointed out that at present the above procedure describing cloning for therapeutic purposes in humans is still theoretical. Several research teams have attempted – unsuccessfully – to reproduce cloning by nuclear transfer in non-human primates (macaque and rhesus monkeys). The cloned embryos obtained degenerated very quickly after only a few divisions; for reasons not yet understood a number of chromosome anomalies developed. These negative results would appear to indicate that any attempts to clone humans today, given the current state of knowledge, are very unlikely to be successful. However, there is no reason to think that this process of asexual reproduction could not one day also be applied to primates, both human and non-human. In all probability, the technical obstacles will gradually be overcome, as they have, in turn, with sheep, cows, mice, goats, pigs, cats and rabbits.

The moral questions raised by this therapeutic use of human cloning relate to how the embryo is viewed. There is a range of opinions depending on the individual and cultural and religious traditions. The cloning of human embryos as a source of cell transplants is undoubtedly tantamount to turning the embryo into a product and as such it is no longer seen as a human being in the earliest stages of development. The ethical debate on this technique therefore revolves around the extent to which the embryo is recognised as a human being and whether a cloned embryo should be regarded as a "potential human being", to use the definition of the human embryo proposed by the French National Advisory Committee on Ethical Issues (Opinion No. 3, 1993).

The definition of an embryo is unambiguous: it is an organism in the process of development from its single-cell stage until the point where it is capable of independent life. In humans, it

Parthenogenetic:
produced by asexual reproduction from a female germ cell alone.

Androgenic:
produced by asexual reproduction from a male germ cell alone.

is customary to use the term "foetus" to refer to an embryo that is more than three months old. The way in which the embryo has been created has no bearing on the definition: alongside embryos formed by the fertilisation of a female egg by sperm from the male, there are also parthenogenetic* and androgenic* embryos, in other words embryos which are of exclusively female or male origin; in plants, the production of embryos can be totally asexual. Since the late 1980s, all the research teams that have successfully cloned mammals by the process of nuclear transfer have referred to the creation of embryos by this process. For humans, therefore, it seems clear that we must use the term "embryo" to describe any stage of development which is likely to continue under the right conditions (that is, in a woman's womb) and without outside intervention until it becomes a foetus and, eventually, a baby. Accordingly, it may be possible to obtain a human embryo by replacing the nucleus of an oocyte by the nucleus of a somatic cell if it can be shown that it could develop in the womb into a foetus and a baby.

The status of the embryo is a highly polemic issue and it is unlikely that any consensus will be reached. And yet, the adjective "human" is used to describe this type of embryo – we refer to a human embryo and not a simian or murine embryo. In a number of cases, this embryo will develop into a child. At the very least, there is the possibility of a person coming into being. It might therefore be useful to consider the matter in other terms: human beings, which the embryo has the potential to become, are due unlimited dignity. What dignity does this confer upon the embryo? French law has refused to confine the embryo and humans within biological definitions and precise time limits. The United Kingdom legislation has taken a different approach by adopting a legal definition of the point in time at which embryo research becomes, in principle, a criminal offence.

Since 1990 (Human Fertilisation and Embryology Act), the UK Parliament has given authoristion for embryos to be created by traditional means exclusively for research purposes in certain key and promising areas of studies into infertility and genetic diseases.

Licences are issued on a case-by-case basis by a special agency (the Human Fertilisation and Embryology Authority) on condition that the embryos are destroyed no later than the fourteenth day of the growth cycle. According to UK law, there is a fundamental difference between the nature of the embryo before and after that point – that is, before the primitive streak of the embryo appears – justifying its use for research purposes and its destruction. Some people have interpreted this legislation as acknowledging the existence of a "pre-embryonic stage" during which research may lawfully be carried out. The legislation is not aimed at transposing a biological fact into law but reflects the utilitarian and pragmatic approach to moral philosophy in the UK: since there is no doubt about the usefulness of research on the human embryo but given that a whole corpus of law protects the human person, making the status of the embryo uncertain, the UK Parliament got round the problem by creating a "virtual" object, a legal fiction under the protection of which biologists can continue their work. Moreover, the scientific considerations behind this British reasoning themselves derive from the utilitarian approach: it is the nervous system which will enable the foetus and the child to feel pain, avoidance of which is one of the main objectives of utilitarian morality. The newly fertilised egg, the single-cell embryo, will develop gradually, will acquire its form and movements, and the nervous system will develop until birth and beyond. The whole process is therefore a dynamic one. In parallel, the relationship *vis-à-vis* this person also develops, characterised by an increasing level of respect. Accordingly, the human embryo, a potential human being with dignity is never, in my view, a mere object. Nonetheless, neither is it a person in society with all his or her rights.

However, if we accept that a human embryo is not a mere object, we have at the very least to view it as a potential person, that is, an end in itself and not simply a means to another end which has nothing to do with the creation of a human life. It is here that the creation of human embryos outside any reproductive process and solely for research or in order to produce therapeutic material* raises a number of questions. Once we have crossed the moral and legal boundaries prohibiting the

Therapeutic material: will serve to heal or ease the pain of patients.

creation of human embryos with no reproductive aim – and they undoubtedly will be if the prospects for therapeutic cloning live up to expectations – what does this mean for the contention that a human embryo cannot be instrumentalised?

To avoid this problem, some people have proposed a solution similar to that adopted by the UK legislation: to take the view that a cloned embryo would not really be a "human embryo", either because, based on the notion of pre-embryo, it would be destroyed before fourteen days, or because it would not have been the result of procreation. This last proposal is odd because, as we have seen, the definition of an embryo, in biology, has nothing to do with how it is produced. Moreover, if a cloned embryo were ultimately to be used to lead to the birth of a child, that person would be unique since he or she would not have developed from an embryo.

A further argument put forward for dehumanising the human embryo is that it is the intention which leads to the project (Frydman and Khan 1999), which is true enough, and the project itself or ultimate aim which gives rise to dignity. Following this line of reasoning, an embryo *in vitro* conceived for the purposes of research should not be considered as a "potential human being". Similarly, the voluntary termination of an unwanted pregnancy ceases to pose the slightest ethical problem. In short, an embryo of *Homo sapiens*, conceived for research or for producing therapeutic cell material without any reproductive aim, is not human and any questions as to how much dignity should be conferred upon it are irrelevant. The UK Parliament probably recognised the flaw in that argument when it set the 14-day limit, after which time the human embryo had to be afforded protection in all cases: the argument of intent could justify the killing of an unwanted child or the premature inducing of labour in order to obtain foetal organs. Peter Singer, an Australian philosopher of Austrian origin, currently Chair of Bioethics at Princeton University in the United States, renowned champion of animal rights and author of a controversial work on practical ethics (Singer 1997), sees nothing ethically reprehensible in the above. He says:

> "... for on any fair comparison of morally relevant characteristics, like rationality, self-consciousness, awareness, autonomy,

pleasure and pain, and so on, the calf, the pig and the much-derided chicken come out well ahead of the foetus at any stage of pregnancy – while if we make the comparison with a foetus of less than three months, a fish would show more signs of consciousness".

All the same, such an extreme view is shared only by a minority; however, in order to avoid such barbarism, those who hold that the human dimension of the embryo results exclusively from a reproductive process are obliged to incorporate a time limit and therefore reintroduce the concept of pre-embryo (even if this is highly debatable in terms of its biological meaning and ethical consequences). Lastly, how can we merely accept that the nature of things and beings depends solely on the good will of an individual? When in Corneille's[1] play *Horace*, the eponymous hero says to Curtius, "And this superfluous language to give o'er, you are Alba's choice, nor must I know you more", he denies Curtius' existence and in so doing makes reality subordinate to his own will. But Curtius replies, "Yet to my torment, I must still know you", thereby demonstrating one of the most obvious and binding of ethical requirements: it is by acknowledging and not denying conflict that the basis for a solution can be found, which takes due account of the nature of events, things and people.

Accordingly, there can be little justification for refusing to give an embryo cloned by the transfer of a nucleus into a female egg the title of "human embryo" simply to remove some of the obstacles to its being used other than for reproductive purposes. This view is not intended to block all roads to a possible therapeutic use which could come about if its potential advantages were confirmed and made available to citizens of other countries. In such cases it would be difficult to see how national parliaments could refuse patients in their countries the benefits of a major innovation in treatment available elsewhere in the world. However, overcoming ethical conflicts, that is, the contradictions between two moral standards with diverging consequences, does not mean that they should be ignored. If therapeutic cloning were ultimately to become established practice, it would be preferable that it should do so because the conclusion reached was that the expected benefits

1.
Corneille, Pierre (1606-1684), was a French classical tragic dramatist, whose plays include: *Le Cid* and *Horace*.

for patients justified instrumentalising to a certain extent "potential human beings" rather than artificially stripping them of their potential to be human beings. Additionally, acknowledging the ethical difficulty of such developments is also another way of emphasising the value of research aimed at providing patients with the same type of treatment as with therapeutic cloning but without going down the road of creating human embryos specifically for this purpose. There are other available options using either surplus embryos that would otherwise be destroyed or perfecting techniques for using adult stem cells which are found in each and every one of us.

A further ethical challenge raised by therapeutic cloning is that the first stage – the creation of human embryos – is obviously identical to the first stage in the process leading to the birth of children. Once thousands of such embryos have been obtained in order to prepare cell populations, it would just be a matter of transplanting some of them into wombs to lead, perhaps, to the birth of cloned babies. This is a crucial consideration at the present time, when we have heard many declarations of intent to carry out human reproductive cloning. The Raelian sect has founded Clonaid, a biotechnological company working on this project. A group of eminent reproduction biologists led by Severino Antinori of Italy has also announced that it has been commissioned by two hundred infertile couples to produce cloned babies using cells from infertile fathers. Both undertakings have the wherewithal to succeed.[1] Sects such as the Raelians, which employ coercive measures, are able to arrange for hundreds of young female "volunteers" to donate eggs and allow their wombs to be used for the transfer of cloned embryos. Antinori and his colleagues have established clinical experience in reproductive biology. The only obstacle facing these would-be cloners is that in all probability they are at present, as we have seen, unable to produce normal human embryos. When the day comes when the technology has been perfected for the purposes of therapeutic cloning, this obstacle will have been eliminated and it should not be long thereafter before it is announced that there are pregnant women carrying cloned foetuses.[1]

1.
See article "Woman pregnant with clone" : http://news.bbc.co.uk/hi/english/sci/tech/newsid_1913000/1913718.stm

A third area of concern relates to the danger of a further instrumentalisation of women's bodies, and no doubt wide use would be made of them for therapeutic cloning. Publicly or privately funded teams carrying out these experiments would have to have a large number of human ova. As demand would undoubtedly create a market, at least in many countries, it is likely that women in need would come forward in droves to join the ranks of paid egg-donors. They would agree, under contract, to be subjected to repeated ovarian stimulation, with the necessary health checks to monitor the quality of the producers and their production. In short, if therapeutic cloning one day proved to be effective, it would be a perfect illustration of one of the conflicts constantly at the heart of ethical debate. In the Aristotelian tradition, ethics focuses on the morality of action to resolve uncertainties and conflicts raised by the application of techniques derived from new knowledge to humans and their environment. As the situations we are contemplating have never occurred before, the solution cannot be provided by professional (ethical), judicial or legislative rules. The main reference used for assessing the legitimacy of action is Kantian: respect for human rights and human dignity, which in this case has to be applied to medical and biotechnological innovations.

Ethical debate often takes the form of resolving a conflict between two moral lines of thought, both of which are legitimate on the basis of acceptable criteria but lead to opposite conclusions. In this case we have the conflict between a desire to ease the distress of patients who could benefit from genuine progress in treatment and the reluctance to instrumentalise the embryo and women's bodies and make it easier for reproductive cloning to take hold. Ultimately, a decision will have to be taken going beyond, but still acknowledging, the ethical questions raised.

Research on stem cells taken from surplus embryos

Not all of the embryos created in the context of medically assisted procreation are transferred into the mother's womb. At present, the future of the embryos not claimed by the couples from whom they have derived remains undecided. Even though legislation in France and numerous other countries throughout the world prohibits all research on embryos which will lead to their destruction, the fact is that these surplus embryos will ultimately be destroyed anyway. In the light of this, can it really be considered that research on such embryos with the consent of the couples involved, as part of a project which has been properly assessed in ethical and technological terms, is an infringement of the respect due to the human nature of the embryo? Recognition of human dignity has never been an insurmountable obstacle to biomedical research on humans of all ages: children, adults or the elderly. Admittedly, the unique factor about research on embryos is that it generally results in their destruction, making it markedly different from other forms of research on humans. However, this objection is difficult to sustain when faced with the fact that the embryos are to be destroyed anyway whether or not there is any research programme.

How would one be showing more respect to a human embryo in destroying it by simply allowing it to thaw out than by carrying out high-quality research which it is hoped would increase our knowledge of and means of overcoming infertility or development disorders? In my view, using embryos, which are not intended to develop into foetuses and then babies, to improve the conditions of other human lives in the future is similar in nature to the transplant of organs from deceased donors where the departed give a living person in need the chance to survive. Surplus embryos which have not been used by the couples from whom they derived and have not been given to other couples will not develop and will therefore never be involved in anything connected with human life, except possibly as part of a therapeutic research programme. The loss of human embryos is not restricted to *in vitro* fertilisation. In nature, eight out of ten fertilised embryos fail to develop and are eliminated.

In short, there is no conflict between the sense of the individuality of the human embryo and the use of embryos that would otherwise be destroyed in research projects which are of high scientific and moral quality.

In conclusion, the prospects opened up by pluripotent human embryonic cells are considerable, in terms of both knowledge and medical applications. However, it is only fair and honest to say that on the medical level, this approach is still only a hope yet to be confirmed.

For the future, therefore, it is a matter of evaluating the actual therapeutic scope for using embryonic cells and deciding on the moral legitimacy of the various techniques involved. A conscious and informed decision will then have to be taken which goes beyond, but still acknowledges, the ethical questions raised.

Bibliography

Bunting, K.D., Sangster, M.Y., Ihle, J.N., Sorrentino, B.P., "Restoration of lymphocyte function in Janus kinase 3 – deficient mice by retroviral – mediated gene transfer", in *Nature Med.*, No. 4, 1998, pp. 58-64.

Clarke, D.L., Johansson, C.B., Wilbertz, J., Veress, B., Nilsson, E., Karlström, H., Lendahl, U., Freson, J., "Generalized potential of adult neural stem cells", in *Science*, No. 228, 2000, pp. 1660-1663.

Comité national d'éthique, "Avis n° 3 du Comité national d'éthique du 23 octobre 1984," in *Les avis de 1983 à 1993*, Comité consultatif national d'éthique, Paris, 1993.

Doetsch, F., Caillé, I., Lim, D.A., Garcia-Verdugo, J.M., Alvarez-Buylla, A., "Subventricular zone astrocytes are neural stem cells in the adult mammalian brain", in *Cell*, No. 97, 1999, pp. 703-716.

Ferrari, G., Gusella-De Angelis, G., Coletta, M., Paolucci, E., Stormaiuolo, A., Cossu, G., Maxilio, F., "Muscle regeneration by bone-marrow-derived myogenic progenitors", in *Science*, No. 279, 1998, pp. 1528-1530.

Frydman, R. et Kahn, A., "Réflexion sur un grumeau de cellules", in *Le Monde des débats*, juin 1999, pp. 12-13.

Goudl, E., Reeves, A. J., Graziano, M.S.A., Gross, C.G., "Neurogenesis in the neocortex of adult primates", in *Science*, No. 286, 1999, pp. 548-552.

Gussoni, E., Soneoka, Y., Stricklan, C.D., Buzney, E.A., Khan, M.K., Flint, A.F., Kunkel, L.M., Mulligan, R.C., "Dystrophin expression in the mdx mouse restored by stem cell transplantation", in *Nature*, No. 401, 1999, pp. 390-394.

Mignon, A., Guidotti, J.E., Mitchell, C., Fabre, M., Wernet, A., de la Coste, A., Soubrane, H., Gilgenkrantz, H. et Kahn, A., "Selective repopulation of normal mouse liver by Fas/CD 95-resistant hepatocytes", in *Nature Med.*, No. 4, 1998, pp. 1185-1188.

Petersen, B.E., Bowen, W.C., Patrene, K.D., Mars, W.M., Sullivan, A.K., Murase, N., Boggs, S.S., Greenberger, J.S., Goff, J.P., "Bone marrow as potential source of hepatic oval cells", in *Science,* No. 284, 1999, pp. 1668-1170.

Pittenger, M.F., Mackay, A., Beck, S.C., Jaiswal, R.K., Douglas, R., Mosca, J.D., Moorman, M.A., Simonetti, D.W., Craig, S., Marshak, D.R., "Multilineage potential of adult human mesenchymal stem cells", in *Science,* No. 284, 1999, pp. 143-147.

Sauter, B.H., Martinet, O., Zhang, W.J., Meteli, J., Woos L.C., "Adenovirus-mediated gene transfer of endostatin *in vivo* results in high level of transgene expression and inhibition of tumor growth and metastases", in *Proc. Nat. Acad. Sci. USA,* No. 97, 2000, pp. 4802-4807.

Shamblott, M.J., Axelman, J., Wang, S., Bugg, E.M., Littlefield, J.W., Donovan, P.J., Blumenthal, P.D., Huggins, G.R., Gearhart, J.D., "Derivation of pluripotent stem cells from cultured human primordial germ cells", in *Proc. Nat. Acad. Sci. USA*, No. 95, 1998, pp. 13726-13731.

Singer, P., *Questions d'éthique pratique*, Bayard Editions, Paris, 1997.

Solter, D., Gearhart, J., "Putting stem cells to work", in *Science,* No. 283, 1999, pp. 1468-1470.

Thomson, J.A., Itskovits Eldor, J., Shapiro, S.S., Waknitz, M.A., Swiergiel, J.J., Marshall, V.S., Jones, J.M., "Embryonic stem cell lines derived from human blastocysts", in *Science,* No. 282, 1998, pp. 1145-1147.

Ethics, morality and religion

by Dietmar Mieth

The step from the cloned apple to the cloned sheep is greater than the step from the cloned sheep to the cloned human being. Human cells have already been cloned, but these experiments were instances of embryo splitting, involving divided totipotent cells*. This procedure is also known from animal experimentation and animal breeding. Researchers in Pittsburgh, Pennsylvania, USA caused a sensation when they used it on humans in 1993. In Germany, however, section 6 of the Embryo Protection Act 1990 *(Embryonenschutzgesetz)* directly prohibits the cloning of embryos and reinforces this ban, as it does the ban on transferring cloned embryos to women, by means of penalties.

At the time when Germany's Embryo Protection Act was being drawn up there was still doubt about the practical feasibility of transplanting the somatic cell nucleus* into a previously enucleated egg, because scientists had not yet been successful in "reactivating the 'deactivated' DNA fragments of the transplanted somatic cells" in mammals (in the words of the commentary on the German Embryo Protection Act, Keller/Günther/Kaiser, 1992). This is now possible, as in the case of the cloned sheep Dolly, where a somatic cell was successfully implanted into an enucleated egg[1].

When this event was made public through the renowned journal *Science*[2] *(New Scientist,* 1997),[3] scientists, politicians and ethicists emphasised that there was a broad consensus that the cloning of human beings should not be allowed and that it should be ruled out by national and international bans or conventions. One definite, though not fully discussed, means of recourse seemed to be the Council of Europe Convention on Human Rights and Biomedicine,[4] with its prohibition of the intentional transmission of genetically manipulated genotypes from one generation to the next. Furthermore, in the journal *Science (New Scientist,* 1997) the argument was immediately advanced that human cloning required a moratorium, but not a ban.

Totipotent cell: a cell that has the potential to become any tissue in the final organism, split into two or more parts.

Somatic cell nucleus: the nucleus of any body cell that is not part of the germ line.

1. See "Some key concepts", Appendix I.

2. *Science*: http://www.sciencemag.org/contents-by-date.0.shtml

3. *New Scientist*: http://www.newscientist.com/

4. Full title: The Convention for the Protection of Human Rights and Dignity of the Human Being with regard to the Application of Biology and Medicine: Convention on Human Rights and Biomedicine http://conventions.coe.int/treaty/en/Treaties/Html/164.htm

Germ line gene therapy:
transfer of genes in the sex cells in order to genetically modify future generations.

Monogenetic hereditary disease:
an inheritable disease resulting from a mutation in a single gene.

The opposing positions are clear, and they are similar to those in the debate on germ line gene therapy.* Here, too, there are those scientists, but also ethicists, who assert that if the method can be used precisely, safely and without any side-effects, with the goal of eradicating monogenetic hereditary diseases*, one is morally obliged to make use of it to spare entire generations from immeasurable suffering. Others, however, want to do without any such manipulation, above all because of the uncertainty of its consequences in a still unknown future context, and also because it is impossible to consult those who will themselves be affected in the future, and, finally, because the excessive use of embryos in experimentation – which is the only way, so far, of achieving germ line transfer – has been rejected. In the light of this controversy, the European Commission Group of Advisers on the Ethical Implications of Biotechnology (GAEIB), in its position paper on gene therapy (1994), cautiously recommended that germ line therapy should not be allowed, on technical as well as ethical grounds, "at the present time".

In any event, it is clear that the ethical discussion on the cloning of human beings is not limited to arguments in favour of a strict prohibition as a matter of principle. With new technical or technological possibilities, a previous taboo" often disappears. By "taboo" I mean a prohibition that has not been analysed and whose advocates are seldom under pressure to justify it. Examples are *in vitro* fertilisation with embryo transfer and prenatal diagnosis. Here the loss of embryos and foetuses has become routine. The methods are selective and continue to be ethically problematic, even though they are permitted by law.

Three positions on human cloning

On closer scrutiny there are three positions as regards human cloning:

1. an abstract endorsement, subject to almost unattainable conditions, so that this option actually constitutes a moratorium;

2. rejection, the aim being to defend the everyday world against radical intervention;

3. an equally strict rejection that is justified not only by irrefutable arguments but also by a series of relative arguments.

Abstract endorsement

The prerequisites for abstract endorsement are:

- ambitious goals compatible with human dignity;
- precise and safe implementation;
- no negative effects on the present or future world.

These conditions seem to me desirable in the abstract but not attainable in practice. Another objection is that the goal of replacing diseased or defective organs with others from a warehouse for human spare parts is just as incompatible with human dignity as the goal of replacing a dead child with an exact copy. The principle of human dignity as formulated by Kant is that the human being should never be treated only as a means but always as an end in himself or herself as well. This state of being an end in oneself would be lost in acts of procreation for instrumental purposes. The desire for another child is understandable; a child copied on request is by definition a replacement.

Equally problematic is the goal of permanently controlling genetic diseases, not through the germ line but through cloning. The uncertainty about the consequences for the environment as well as the fact that there is no possibility of obtaining the consent of the human beings who will be affected by this new world are the crucial arguments here. One suspects that the position based on a conditional endorsement, currently under negotiation, has been adopted for strategic purposes: if one agrees to it, the next question is automatically and immediately: "What degree of risk is one willing to accept in practice after having already agreed *in abstracto*?". I consider this position to be not so much an attempt to justify cloning on ethical grounds as an attempt to outwit the inexperienced participants in the discussion. Or, to use a metaphor appropriate to the context of Dolly, it is an attempt to pull the wool over the other participants' eyes. This is nothing new: present or future problems are cast aside when the question of the permissibility of cloning is debated. Once consent has been obtained, the dossier is reopened, and one struggles to little avail with the problems. In my opinion, the principle here is that problems should not be solved in such a way that the resulting problems are greater than the initial problem.

With human cloning, potential scenarios for everyday life emerge that are truly shocking, especially if we consider our morality a product of our everyday world, that is the world of our daily experiences, in which we become individuals in a social context, our life in the community as a communication process. Science and technology come into it, but they must remain integrated in this everyday world. When "creation" or "nature" is mentioned by religious individuals or by those with a "religion as a heritage" (Ernst Bloch's "*Religion im Erbe*") in this context, these symbolic words indicate a stable point of reference in this everyday world. We could also speak of the achievements of human history and culture. These include our culture of sexuality and the situation of human reproduction in identifiable relationships. The view that sexuality, because it is totally separated from reproduction, will one day be reduced to repeatable brain stimuli is just as great a culture shock as the view that human beings will one day deliberately be assembled from different "biological" materials.

The human being reacts to the shock of threats to our every-day world with taboos. Taboos offer a sense of protection, and act as a wall against anything that appears to threaten what is vital to everyday life. "Taboo" has, of course, become an ambiguous word, sometimes even an openly abusive word, but the rationality of apparently irrational conceptions (irrational because they are not spontaneously justifiable) can often be shown when they are challenged. Hans Jonas's "hermeneutics of fear" *("Hermeneutik der Furcht")* – when it comes to options for future worlds, one should sooner pay attention to that which is to be feared than that which is to be hoped for – is, in my opinion, nothing other than proof that negative experiences lie behind taboos, for example the memory of the eugenics policy of the Third Reich.

Rejection of cloning

The second approach – the rejection of cloning – seeks to protect our everyday lives from change and upholds conservative values. Spontaneous statements by politicians frequently reflect this consensus on conservative values. But how long will it hold? How long will our everyday world, which is constantly changing, remain a stable source of resistance? Will politicians not find it easier to limit the most drastic examples of abuse, on the one hand, and allow each individual to make use of technological progress as he or she chooses, on the other? The individual option is always the choice of the decision makers, to whose views developing or future human life will have to conform. In the face of such prospects, which have been borne out only too well by the experiences of the last decades, we must attempt to come up with a justification for the rejection of human cloning that is not merely based on a single irrefutable argument but instead resembles a cable, in which several arguments, which are possibly of relative importance if taken alone, are combined to provide resistance.

Rejection based on a series of relative arguments

This is the third approach, which I will defend below: a series of relative arguments that act together to support a strict prohibition.

The first relative argument arises from the weaknesses of the conditionally permissive position, which I have already described.

A second argument can be described as the non-instrumentalisation argument. According to this argument, each human being, even if he or she does not have the right to his or her own unique identity, has the right not to be constructed as a copy of another human being as a result of a third person's plans. With cloning it is, of course, conceivable (provided the "[surrogate] mother animal" brings the foetus to term) that no perfect copy, that is to say no totally identical copy, will be produced, because the cells may adapt differently, and because the environment can also have a modifying effect. Nonetheless, I think that in acting as responsible human beings – "nature" bears no responsibility, for example, for the creation of identical twins – we must seek to ensure that everything possible is done to achieve a unique identity. Even if "identity" is not an irrefutable argument against human cloning, the non-instrumentalisation argument is irrefutable in most of the cases which we can imagine.

Thirdly, the road to Dolly was paved with foetal experiments. We must ask ourselves the moral question – which already applies to animals, although not in such categorical terms – of whether we can take responsibility for experimentation involving such an excessive number of foetuses, or in general for a method entailing trial and error to such a large degree.

The fourth argument is analagous to the prohibition of germ line transfer (on the basis of the uncertainty surrounding it and the absence of the informed consent of the human being affected under changed circumstances). Here it must be added that, as with germ line transfer, the borderline between illness and "design on request" has become blurred. If illness (to use the World Health Organisation's definition) is the impairment of physical, psychological and social well-being, then the borderline between therapy and the genetic improvement of the human being can, subjectively speaking, be quickly crossed.

This bring us to the fifth argument, which is that every application, no matter how limited initially, lands on a slippery slope. Who will be able to determine the individual boundaries once the borderline as a whole has been abandoned? It is also, in relative terms, an ethical argument, in which we weigh up the alternatives, to decide where the line is best drawn and maintained.

The sixth argument is the uncertainty surrounding our everyday world. Can one uphold the family as the nucleus" of society and, at the same time, risk undermining its structure, even if apparently only marginally for the time being? The idea that women will one day donate the eggs, and men will warm the incubators, with cloned designer children under their skirts, may seem satirical and abstruse, but perhaps the day will come when someone comes up with the idea that this is not without precedent in nature: for example, the male albatross cares for the brood. Are there certain things that we do not want to give up? Are there constraints and pain and suffering that we are willing to continue to accept because the price to be paid for their elimination is too high? These are serious questions, and it is our moral responsibility to find answers to them.

To return to my metaphor, these arguments are like a cable formed from six different strands. Individually, the arguments may be of relatively minor importance, except for the non-instrumentalisation argument, in most of the cases imaginable. The individual arguments cannot totally dismantle every counter-argument that could be imagined. Nevertheless, they carry weight and thus strengthen the position in favour of a ban on cloning. A wall built of individual stones can be as sturdy as a rock. Yet the individual arguments are open to discussion when assumptions, contexts and consequences change. Technical possibilities can and will challenge ethical reflection but cannot replace it. As long as we understand our culture as the sum total of life in society, our everyday world cannot be determined – and we cannot allow it to be determined – one-sidedly by an alliance of science, technology and business. Therefore I would argue for not only a moratorium but a ban on human cloning.

*In vivo** cloning and *in vitro** cloning

Our language is not capable of grasping all new phenomena immediately. Any concept therefore has the potential to be misunderstood and this applies to the distinction between "reproductive" and "non-reproductive" cloning. Reproductive cloning means that the clone is implanted and a human adult develops from it. The term *in vitro* does not allow us to determine whether an early embryo that has been manipulated will or will not be implanted. If it is not implanted, experiments may be carried out on it with particular long-term therapeutic goals; this counts as "non-reproductive cloning".

The EU advisory group[1] came to the conclusion that "reproductive" cloning must be forbidden. President Clinton's National Bioethics Advisory Committee,[2] using the same terminology, was more liberal insofar as it demanded a moratorium for the time being. The phrases "for the time being" and "in the present social context" did not appear in the European advisory group's conclusions. In my opinion, these qualifying phrases

In vivo:
occurring in a living organism.

In vitro:
biological processes or reactions occurring outside a living organism e.g. cells or tissues grown in culture.

1.
EU European advisory group: European Commission Group of Advisers on the Ethical Implications of Bio-technology (GAEIB) 1991-1997, succeeded by the European Group on Ethics in Science and New Technologies (EGE); http://europa.eu.int/comm/european_group_ethics/index_en.htm

2.
Although the Charter of the NBAC expired on 3 October 2001, information on its reports, findings and meetings can be accessed at: http://bioethics.georgetown.edu/nbac/

imply that any such prohibition will automatically be reviewed! It is self-evident that every rule that we formulate under present circumstances may be reviewed in the future. But anyone who wants to use these particular expressions has ulterior motives that must be put on the table. Thus it can be said that the Clinton Bioethics Commission declared itself in favour of a moratorium, whereas the European advisory group was in favour of a strict ban, but only in the so-called "repro-ductive" area.

In the "non-reproductive" area, that is, with reference to *in vitro* cultivation of cloned embryos in cell cultures, the EU advisory group stated that in those countries which allowed experimentation on embryos (such as Belgium and the United Kingdom), *in vitro* cloning should not be forbidden either, on condition that it was carried out for high-priority therapeutic purposes, and on condition that a licensing body, that is, an ethics commission was consulted and, finally, on condition that the manipulated embryos would not be implanted and become independent human beings. In my opinion, the distinc-tion between "reproductive" and "non-reproductive" cloning was not made on a factual basis; rather, the aim was to differen-tiate between the treatment of "embryos" and "human beings".

In a 1994 document I found a statement by the Committee of Ministers of the Council of Europe regarding tissue banks, which indicated how the terms "reproductive" and "non-reproductive" were to be understood in this area. Egg cells, sperm cells and embryos were considered to be "reproductive". Their use as tissue was to be forbidden precisely because they were "reproduc-tive", that is, because they could engender human beings. The expression "reproductive" was therefore intentionally narrowed down by the EU advisory group, in the interests of practicabil-ity, to mean only implanted embryos. I call this the politics of language. The distinction between "reproductive" and "non-reproductive" is not without precedent; it was previously used by an American advisory group on *in vitro* fertilisation.

Non-reproductive cloning is now called "therapeutic cloning"; this is another example of the politics of language. The concept of therapy or the concept of human health plays a central role in the politics of language. I would like to elucidate this with an example from the Council of Europe Convention on Human Rights and Biomedicine. Article 18.2 of the convention states that the production "of embryos for research purposes" is forbidden. Article 14 deals implicitly with research on embryos, although explicitly it is concerned with the possibility of sex selection. It states that this is possible only for health purposes. As an example of health purposes, sex selection in order to prevent hereditary disease is mentioned, since a hereditary disease can be connected to a particular gender. The text does not clarify which method it is referring to: abortion after prenatal diagnosis, embryo selection before implantation, or possibly also sex selection through semen centrifuging*. At the same time, Article 12 states that these health purposes also include "research linked to health purposes". Of course, one wonders at this point what is meant by the prohibition of the creation of embryos for research purposes in Article 18.2. Is research for health purposes, which elsewhere is always allowed, ruled out? If this were the case, then the research purposes referred to in Article 18.2 would mean little, because one could claim that any research purpose in this area has a therapeutic purpose.

> **Semen centrifuging:**
> the spinning of semen in a centrifuge (in a two-phase aqueous solution) which allows the separation of sperm enriched in either the X or Y chromosome, thus aiding sex determination of offspring.

The European Convention on Human Rights and Biomedicine and the Additional Protocol on the Prohibition of Cloning Human Beings, which was signed in January 1998, leave the establishment of what constitutes a human being to member states. In an interview with the German weekly journal *Focus*,[1] the chair of our EU advisory group, Noëlle Lenoir, explained that "when human life begins is determined by the nation-states". As one can imagine, when attempts are made to reach a valid European consensus, the fact that this question is left open means that any consensus remains unclear in the specific case under consideration. According to Noëlle Lenoir, Germans consider an early embryo to be a human being, whereas other countries recognise individual human beings only after birth. The Convention on Human Rights and Biomedicine does not specify what constitutes a human being. This is particularly significant for the cloning of human beings. The Additional Protocol to the convention states that it is forbidden to create

1.
http://www.focus.de

127

a human being with the intention of making him or her identical to a living or deceased person. Anyone reading this without bias will assume that this means that any human cloning is forbidden. Yet things are not so simple, even though Article 2 states that there are no exceptions to this prohibition. When one consults the explanatory memorandum things look different.

The explanatory memorandum states that one has to distinguish between three situations as regards cloning. First, there is the cloning of cells, which does not generally pose ethical problems. I too am of this opinion. Cloning is not morally problematic when we are dealing with a living organism without an independent destiny. Secondly, there is *in vitro* cloning. Here "embryo cells" are carefully considered. But "embryo cells" may also cover embryos in the totipotential state. Thirdly, there is the cloning of "human beings". This means that the cloning of "human beings" is distinguished from the cloning of embryo cells. It is then expressly indicated that the protocol applies only to the cloning of "human beings".

This brings me back to my discussion of the politics of language: although the expression "non-reproductive cloning" is not used here, it is clear that the technical distinction which it embodies – *in vivo* cloning, no; *in vitro* cloning, yes – was adopted by the Convention on Human Rights and Biomedicine. Now, the interpreters of the convention claim that this is not that serious since the aforementioned Article 18.2 states that one may not produce embryos for research purposes, and this rules out the possibility of using germ cells to create embryos. Yet I have already pointed out that the concept of "research purposes" is not yet clear. Is research for health purposes really ruled out by this? In the future the courts will seek to clarify this point, since at present it is not clear from the text. Thus, one cannot assume that the Convention on Human Rights and Biomedicine and the Additional Protocol to it totally prohibit cloning. Such a prohibition does not actually exist at all in the Unesco Declaration on the Human Genome and Human Rights,[1] since it forbids only "reproductive" cloning.

1.
http://www.unesco.org/ibc/en/genome/

The importance of the politics of language becomes clear when one considers that the representative of the Vatican, in consultations on the Unesco declaration, apparently supported the prohibition of "reproductive" cloning strongly, without realising that it was only a partial prohibition. This whole linguistic game is not immediately obvious to non-specialists among theologians, since they do not regularly have to deal with such matters. The last opinion of the German Catholic Bishops on the Ethics of Biotechnology states that "reproductive" cloning should be banned, maintaining the misconception that all kinds of cloning are included and that "reproductive cloning" is only a redundant description of cloning as "reduplication". This explanation was given to me when I asked why. But the Church of Scotland has seen the problem clearly enough: "Can we ethically create embryos by cloning – knowing that on ethical grounds we could never allow them to develop to full term?"

This distinction also plays an important role in the cloning prohibition called for by the World Health Organisation. There, almost identical words forbid the creation of a human being as a copy of another, and this always means *in vivo*. Thus, internationally, there is only one place so far where there may be support for a total prohibition on human cloning. The European Parliament made its position clear in a declaration on human rights on 15 January 1998, in which it expressed concern about the announcement of an American researcher that he intended to clone human beings; human cloning being defined as the creation of human embryos with the same genotype as another distinct living or deceased human being in his or her total development, beginning with conception, irrespective of the method used. The European Parliament again confirmed that each individual had the right to a genetic identity and that the cloning of human beings must be forbidden without exception.

In Unesco, in the Clinton Bioethics Commission and in the Council of Europe, with its Convention on Human Rights and Biomedicine, there is obviously overall agreement that *in vivo* cloning, that is, cloning that results in the creation of human beings as individual copies, should not be allowed. However, that type of cloning is nonetheless under consideration. There is, for example, the suggestion of circumventing male infertility by using the man's somatic cell together with the woman's ovum, thus making children possible for a married couple. Richard Seed advocated this as an extension of existing reproductive medicine. Another option entails parents who have lost a child and who want a "replacement" using one cell from the deceased child in order to clone another one, who is supposed to bring the parents the joy of repetition. This option is already problematic for purely technical reasons, but also for psychological ones. Such intentions were seriously proposed as ethically irrefutable by ethicists on the various commissions. A third example originates from Philip Kitcher, who wrote a philosophical contribution on genetic engineering methods. A child has kidney problems, for example, and the parents breed a second child in order to have a spare kidney, which will not carry the risk of rejection.

The reason for forbidding such options has been the same at every level of political consultation. The ethical principle is the "non-instrumentalisation" of the person. The principle of human dignity according to Kant is: act in such a way that you never regard the other just as a means to an end, but also as an end in himself or herself. The person on the receiving end of our behaviour should always also be an end in himself or herself. This is not a principle forbidding instrumentalisation, but, rather, a principle forbidding instrumentalisation of the "design" of the whole person, because the person's "end-in-him/herself-ness" must be preserved. This means that a person who has to see himself or herself as the copy of another, made at the request of a third party, would as such be subject to instrumentalisation at the request of the third party, and precisely this would apply to all three examples which I have submitted. The request of the third person, in this case the wish of the parents, and this is important, does not only relate to existence. It is not just a matter of a person coming into existence in the sense of wishing for a child of one's own; it is a matter of design: it is a matter of the essence of this person. To put it another way, it is interference with the essence of the future person in such a way that he or she serves purposes which are externally imposed.

At first glance this argument seems very convincing, especially in our tradition. But one must not overlook the fact that there are strong counter-arguments. These have been put forward by P. Kitcher and D. Beyleveld. They recognise the Kantian principle: act in such a way that you never regard the other just as a means to an end, but also as an end in himself or herself. Thus, we are not dealing with an argument against Kant. It has been argued, however, that we are not dealing with total instrumentalisation if a second child is created for organ transplant purposes because this child would not be restricted in its moral capacity, which is, after all, what really goes to make up human dignity. This second child would be free, and have the capacity to act, and have a moral capacity, and as long as the moral capacity of a future human being is not restricted, one can speak only of a form of instrumentalisation that is consistent with human dignity.

But I think it is problematic to simultaneously engender anthropological qualities which are necessary for moral capacity and anthropological qualities which are necessary in order to respect other people. I consider it to be an "ethicistical" fallacy (see D. Mieth, *Moral und Erfahrung II*, 1998) to use anthropological simplifications – which we need in order to reach ethical judgments and to provide a basis for morality – directly for specific practical purposes. The capacity to act, moral capacity, freedom, etc. are such simplifications because this is not how a human being can be conceived of at all stages of life, in all their complexity. I am of the opinion that when we speak of human beings as the object of our recognition and regard, we may not transfer the normative reasons for justifying this recognition directly to the specific practical application, because otherwise it will always be those persons who are somehow defective or not yet fully developed in their capacity for freedom, action and morality who are likely to suffer as a result. Interchanging moral reasoning and the anthropologically determined application of morality seems to me to be wrong.

I therefore uphold the principle of non-instrumentalisation and am of the opinion that it is right to condemn *in vivo* human cloning worldwide. In 1997 we had a once-in-a-lifetime chance to introduce such a prohibition, because of the general alarm.

This prohibition, however, as I have tried to show, is not total. It is an *in vivo* prohibition. It does not include an *in vitro* prohibition, and that raises the question of the extent to which the argument of non-instrumentalisation can be applied *in vitro*. As regards political consultation, a protocol on the protection of embryos to the Council of Europe Convention on Human Rights and Biomedicine has yet to be prepared. Indeed, the explanatory report to the protocol on cloning refers to the non-existent protocol on embryo protection. In particular, the discussion in the German-speaking countries on the Convention on Human Rights and Biomedicine, which is, after all, not concerned only with this problem, depends on such a protocol.

Theological arguments on the ethics of human genetics

If one accepts the concept of an autonomous morality in the context of Christian ethics, as do many German-speaking moral theologians (J. Fuchs, A. Auer, F. Böckle, etc.), this does not necessarily mean a decision in favour of a particular philosophical way of thinking in ethics, but reflects only the insight that the basis for judgments is achieved by philosophical means. The context of the discovery that ethical values and obligations contribute significantly to ethical motivation, that is, to the motivation for ethics, is assigned to theology. Finally, a theological approach to ethics – one need only look at the doctrine of grace and justification – has a very specific function. Theology is relevant to the discovery of ethical problems, to moral awareness, to moral motivation and when it comes to relativising moral judgments on people. In contrast to Kant, the epitome of being human is not moral capacity but rather the need for salvation, the "feeling of absolute dependency", to supplement Kant's anthropology with Schleiermacher's. Happiness, freedom and God are not philosophical postulates, but rather specific religious experiences.

Theology cannot therefore be expected to replace, circumvent or change the ethical arguments in the biomedical and biotechnological debate in general or in the debate about human genetics in particular. This can be illustrated by an example. In the debate on the protection of embryos, the Roman Catholic Church does not labour the point that from the beginning humans are made in the image of God, and thus bearers of dignity, that is, of personhood before God (person and human dignity are interchangeable in Church documents). However, does this doctrine not spare us the need for philosophical reflection in which we must first explain what living entity is meant when speaking of God's image? This could give rise to abstruse speculation. In line with biblicism, for example, the words of the Psalmist "Upon thee was I cast from my birth, and since my mother bore me thou hast been my God" (Psalm 22:10) would mean that being the image of God should be understood as beginning at birth. Or the words of Psalm 51:5, "And in sin did my mother conceive me", could be understood to mean that personhood begins at conception. The Bible cannot be used to answer questions which did not exist in biblical times. It is more likely that its method of discourse can serve as a model for solving controversial problems.

A positivist approach that regards the fertilised egg as a person only for reasons of caution (so-called "tutorism") would not help either. It could, however, be reinforced by philosophical arguments about the continuity of the human being from the very beginning. But any philosophical argument remains vulnerable and hypothetical; it is never definitive.

If ethical rightness is not a matter for theology (except possibly in that tautological form which prohibits murder, where murder is by definition unethical), then the theological argument can "only" address ethical insights in the context of anthropological conditioning. This is ethically relevant without making philosophical methods irrelevant. This does not mean that the significance of theology is in any way played down or diminished. The significance of such contextual arguments can be that they change the whole ethical context and, in so doing, they can significantly influence the relative importance of ethical arguments. I shall now illustrate this.

When asked whether there is a relevant theological argument against the cloning of human beings, I have said, without neglecting the previous explanations, that the diversity of humanity is a commandment from the creation. This is Ellen van Wolde's conclusion in the examination of Genesis 11. The many languages imposed by God are not an answer to human hubris, since towers that reached up to heaven were in no way interpreted as storming heaven. (This is better known to orientalists than preachers!) It was the urban concentration of humanity, its inclination to uniformity in culture (and religion?) that displeased God. He therefore reminded people of the commandment from the creation that they fill the whole Earth, and he dispersed humanity by confusing its tongues. Something similar was repeated at Pentecost: the gospel, communicable despite the many languages, was spread throughout the world. Of course, diversity, rather than repetition of the same, can be applied as a theological argument only if we have good ethical arguments in its favour, and we have these as I have tried to show above. Just as the point about the person being the image of God and the point about dignity are merely a source of inspiration in ethics, and not substitutes for arguments, the diversity consideration is a theological stimulus but not a substitute for arguments.

The stimulating effects of such theological considerations transcend the problem itself, take for example finitude, that is to say, contingency. Contingency is also a philosophical concept. In classical philosophy it means the dependence of being, being time dependent, finite in the sense of ending at death, capable of mistakes, the imperfection of all that is human. This is exactly what is meant by the theological theorem of the creatureliness of humanity – of not being like God. From a theological and philosophical angle – both approaches are combined in Pascal, for example – by trying to transcend its own finiteness, humanity would gamble away the quality which makes it human. Theology expresses this narratively in the story of the Fall. Philosophy draws attention to the fact that it is precisely in the corporeality of being human that both fulfilment and freedom from want are located (for example, according to Merleau-Ponty), that finiteness is the form that the possibility of happiness assumes.

Finiteness could be a concept over which theology and philosophy converge. It is relevant in ethics because it interprets the following approach as missing the point of what it means to be human: the expectation of humanity's perfectibility, of the feasibility of achieving anything, and of still being able to correct all the problems that are created anew by problem-solving. This interpretation considerably changes the context of the biomedical and biotechnical debate. For example, it is grist to the mill of the ethical rule of problem solving: one should not solve problems in such a way that the resulting problems are greater than the problems that are solved. I do not believe that the paradigm of progress in modern biotech-nology, including human genetics, takes account of this con-textual insight. In bio-politics I have observed that in this regard continental Europeans, even if they are agnostics, can be convinced of this more easily than people from the United States, even if they consider themselves to be Christians. How is this possible?

In conclusion I want to introduce a fourth theological argu-ment – after the arguments of being made in the image of God, of the creation commandment to diversify, and of finiteness – which I often painfully experience as missing from the context of reflection: the vulnerability of humanity. By analogy with the theological "option for the poor", in our context (biotech-nology, human genetics) it is possible to formulate an option for the priority of vulnerable persons. Here one is likely to come up against a lack of understanding in the secular context. Certainly, the average pragmatist discussing ethics in political advisory groups will acknowledge that one must respect dis-abled persons and incompetent persons, and even (to a greater or lesser degree) foetuses and embryos. But he or she will react to a theological theorem such as "the prophecy of disabled per-sons" with little understanding. And yet, it is easy to explain what it means: being explicitly disabled reminds us that we are all disabled in some way. It reminds us of our finiteness.

In seeking ethical judgments in an accelerated world where doubt is suspended, there is often a temptation to represent inarticulate people who are not able to defend their interests and express their opinions themselves by extrapolating their presumed intention via analogous deduction. The fiction of "informed consent" is stretched to include them. Curiously, this very process tends to instrumentalise them. When, for example, the Council of Europe Convention on Human Rights and Biomedicine tried to formulate the conditions under which research for the benefit of others could be carried out on those incapable of consent, it forgot to secure the possibility of withdrawing consent given by proxy. A whole mindset can be recognised in this oversight!

There is an aura of vulnerability about people who cannot stand up for themselves, the lustre of the butterfly that may not be touched. Forgive the poetic expressiveness, but I want to illustrate that we are dealing here with the need for greater sensitivity. Sensitivity can easily be declaimed. It is then up to us to prove that it is not present. Nowadays, we live with the ideal of the person as young and strong, an ideal for which Paul admonished the Corinthians. It is not just a matter of developing a culture of perception and feeling (although that is necessary too). It is a question of the role of taboos which have grown out of historical experiences, for example the Nuremberg code, which simply forbids experimentation on those not capable of self-determination. Taboos can be ethically justified only if they are understood as cautionary fences. They do not belong to ethical justification but to its context, which modifies the importance of the justification. In other words, situations of vulnerability demand caution.

The reticence of the Roman Catholic Church in this regard puzzles me. The Church has either not reacted, or reacted with cautious agreement, to the international advisory texts, declarations (Unesco) and conventions (Council of Europe). On the other hand, it disciplines theologians and the faithful for reasons that are difficult to defend. Even the suspicion of ambiguity in pregnancy crisis counselling (obligatory counselling in the three months during which abortion is not penalised), which could be (but in practice is not) the responsibility of Church-related counselling services, has led to vehement interference with the German bishops' actions by the Pope and the Prefect of the Congregation for the Doctrine of the Faith. Not even practices based on considerations of mercy may stain the clean white teaching robe of the Roman Catholic Church, which bans abortion and penalises it by excommunication. On the other hand, Church authorities hardly, or only very indirectly and abstractly, criticise possible abuses of biotechnology, that is to say, human genetics. Should the Roman Catholic Church, in opposition to the intentions of the Pastoral Constitution of Vatican II, move towards a split between the message for the world and the message for the Church? The inclination to pay close attention to ethical problems only when they are ecclesiastical problems exacerbates the retreat into the Roman Catholic ghetto more than would be the case with a well-argued critique of the "modern" system based on the interdependence of research, technology and the economy. However old and wise the Church authorities may consider themselves to be, the discussion on "bioethics" is yet to come; it is not over and done with.

The anthropological question: the ethical status of the early embryo

What is a fertilised egg cell, an early embryo, from an anthropological (philosophical) perspective? The Council of Europe Convention on Human Rights and Biomedicine, as we have seen, offers three possible answers, which are mentioned in the Additional Protocol on the Prohibition of Cloning Human Beings:

- a person;
- a human being;
- and a conglomeration of human cells.

I suggest that the last of these three possibilities be ruled out, because there is indeed a significant difference between human germ cells, that is, egg and sperm cells, and an early embryo. The embryo has a gender! It is a fact that the embryo has the ability – and not merely in the sense of an abstract potential, but rather in the sense of a real capacity – to become a human being, if its development follows the inherent intention, that is to say, when nothing is undertaken or omitted to the detriment of this development. The objection that "nature" does not implant all fertilised egg cells either does not count in this context because "nature" cannot be treated as a morally responsible subject. Anyone who is not prepared to accept "nature" as an ethically restrictive argument should not attempt to use it as an argument for indifference, when it comes to drawing up rules, either.

If, however, when talking of an embryo we are dealing with a human being, then we need at the very least to speak of its having a morally relevant status. Can one maintain, on the basis of this status, that every embryo is the bearer of individual rights which preclude its destruction or even its being put at risk? In my opinion, those who do not want to protect embryos individually, but would rather protect them only as a particular kind of "biological material" which is to be treated with respect, by allowing their use only to a very limited extent, are already violating the morally relevant status of a human being to which I have just referred. My point of view assumes, of course, that belonging to the human species in itself entails a particular right to protection which transcends that which pertains to animals.

But is the issue not broader than this? The Roman Catholic position proceeds from the assumption that an embryo is to be treated "like a person". This wording is carefully chosen, to the extent that it does not simply mean that embryos are identical to persons. It is argued, of course, that, because of the unity of human development and because of its continuity, in that it cannot be broken down into completely different phases, even according to the major stages of development – and indeed because of the unpredictable consequences it would have for other areas if one began to distinguish between human beings on the basis of this kind of personal characteristic – one cannot make a distinction between "human beings" and "persons" and assign them to two different levels. The main argument here is thus the continuity and inseparability of being human. The consequence is that embryos acquire a moral status that confers full protection of life and does not allow research which treats them as raw material. If one upholds this status, then life as a fundamental right cannot be weighed up against other high-ranking considerations.

This position, which is itself philosophically argued – in the absence of revelations in this area which can be proved to transcend a particular time and place, a theological position can only be philosophically argued – can also be attacked on philosophical grounds.

This position also rules out the possibility of a distinction that has become common in the indifferent pragmatism of the sciences concerned, namely that drawn between embryos considered worthwhile, to be used for implantation, and embryos considered inadequate, to be used for research. With this kind of decisionism, one spares oneself any need for further ethical reflection.

There is a tendency among the supporters of research that uses embryos as raw material to make absolutes out of the medical options and the therapeutic benefits, which are only potential benefits and by no means certain. Research, however, uses material *now*, and irretrievably at that. The interests of future sufferers from an illness are important, but they must not be made into absolutes in a society that needs to be committed to all the relevant values. The social solidarity that is reflected in the resources placed at the disposal of research also needs to reflect priorities.

Bibliography

Ach J.S., Brudermüller G., Runtenberg C. (eds.) (1998). *Hello Dolly? Über das Klonen.* Frankfurt a M., Suhrkamp.

Haker H., Beyleveld D. (eds.) (2000) "The status of the embryo", pp.59-100, "Legal regulations", pp. 215-294. In *The Ethics of Genetics in Human Reproduction.* Ashgate publishing, Aldershot UK and Burlington USA

Kaiser G. and Schmidt S. (eds.) (2001). "Stammzellen und therapeutisches Klonen – Biomedizin ohne Grenzen?" *Das Magazin*, 12, No. 2.

Mieth, D., *Moral und Erfahrung II, Grundlagen einer experientiellen Ethik*, Studien zur Theologischen Ethik, Freiburg/ Schweiz-Freiburg i. Br. 1998.

Legal responses

by Maxime Tardu

In the face of revolutionary scientific progress such as the enormous strides made in the field of New Biology, the initial, instinctive reaction of lawyers is to remain silent. Seeing their usual references turned upside down, they bide their time until new references emerge. Hence the many countries where cloning is still a grey area, or even *terra juridica incognita*.

Nonetheless, under the influence of powerful currents of opinion, fuelled in many cases by emotional public reactions, laws and treaties can develop very quickly before any proper legal debate has taken place.

Some discoveries have derived benefit from this, the enthusiasm for compulsory vaccinations being a case in point. However, the strong reactions to human cloning at the moment are mostly negative.

Since the birth of Dolly in 1997 a number of countries have speeded up the process to enact legislation outlawing cloning. In some cases, the bans extend even to research for therapeutic purposes.

At international level, meanwhile, the binding 1998 Additional Protocol to the Council of Europe's Convention on Human Rights and Biomedicine[1] explicitly prohibits reproductive cloning. A number of international declarations, and in particular the 1991 Unesco Universal Declaration on the Human Genome and Human Rights,[2] either echo the Council of Europe convention or even broaden the scope of the ban.

While it is true that there are some multilateral laws and recommendations which make a distinction between reproductive cloning* and therapeutic cloning*, banning the former while supporting the latter, such texts are still few and far between. Similarly, still only a minority of countries have adopted the moratorium approach.

Reproductive cloning: production of a human being that is genetically identical to another (by nuclear substitution from a human adult somatic cell or child cell, or by artificial embryo splitting).

Therapeutic cloning: cloning where the object is not to implant the clone, but where experiments may be carried out on it with particular long-term therapeutic goals, or where it may be used to grow tissues for therapeutic transplantation.

1.
Full title: The Convention for the Protection of Human Rights and Dignity of the Human Being with regard to the Application of Biology and Medicine. See also Appendix III. See: http://book.coe.fr/conv/en/ui/ctrl/menu_en.htm

2.
http://www.unesco.org/ibc/en/genome/

In the first part of this chapter, I shall attempt to define the different types of national laws. The second part looks at international law in relation to cloning.

Different types of domestic legislation on human cloning

Total ban

A sizeable proportion of the laws that refer explicitly to cloning appear to impose a total ban on all forms of cloning, reproductive and therapeutic, and in some cases even research into cloning. In Germany, for example, Section 6 of the Embryo Protection Act of 13 December 1990[1] provides for criminal sanctions of up to five years' imprisonment. In Switzerland, Article 119(2)(a) of the new Constitution in force since 1 January 2000[2] provides that:

> "(t)he Confederation shall legislate on the use of human reproductive and genetic material. .... in particular it shall respect the following principles:
>
> a. All forms of cloning and interference with genetic material of human reproductive cells and embryos are prohibited."

A bill prohibiting all forms of cloning is currently being drawn up in Canada. In some countries, cloning would effectively already be banned under existing rules and regulations. This is the case in Greece, for example, by virtue of general legislation and the Greek Constitution. The protection afforded to "the unborn" under the Irish Constitution would also effectively prohibit cloning.

In France, following the publication of a report by the National Advisory Committee on Ethics of 22 April 1997, the feeling in some circles was that there was no need for new legislation since cloning would effectively be extensively prohibited under provisions of the Public Health Code when taken together with provisions of the Civil Code. However, the 1998 bioethics laws, which contained no provisions explicitly prohibiting human cloning, are currently being revised.

1.
Bundesgesetzblatt, Part I, 19 December 1990.

2.
Recueil officiel des lois fédérales, 26 October 1999, No. 42.

Legal situation in the United States of America: from freedom to prohibition?

In the USA, the bill voted on in the Senate following its adoption in August 2001 by the House of Representatives bans not only reproductive cloning but also all creation of embryos for therapeutic purposes.[1] A number of states, including California, have already passed laws banning or restricting cloning.

Earlier, the US Food and Drug Administration (FDA) had tried to use the Public Health Service Act, which gave the Administration jurisdiction to regulate all "biological products", to assert its authority to ban cloning,[2] but many constitutional experts rejected this very broad interpretation of the law. In any event, the Food and Drug Administration should confine itself to assessing the effectiveness and impact of cloning for the health of the individual concerned and should refrain from making ethical judgments.

The crucial question is what attitude the American Supreme Court would take towards federal and state laws banning cloning. American lawyers and experts in the ethics field who are opposed to the radical ban on cloning point out that the Supreme Court has a long tradition of adopting a liberal approach to the choice of "procreation" method, and freedom to procreate has been considered by the Supreme Court to be a "fundamental right", notably under Amendment XIV[3] ("Due Process Clause") to the American Constitution. Is cloning still "procreation", however, or is it "reproduction", in the strict sense of the term, or "replication"? Would the Supreme Court extend the scope of its protection to include cloning, a non-coital, non-sexual method of reproduction?

Ethics-related concepts used to support anti-cloning laws

Since other chapters appearing in this publication examine the great ethics debate on cloning, I shall merely attempt to highlight the main ethics-related concepts underlying anti-cloning laws. Reference should be made not only to actual laws but also, wherever possible, to parliamentary debates, to reports by committees of experts, which are nearly always consulted, and

1.
International Herald Tribune, 16 August 2001.

2.
Washington Post, 23 May 2001.

3.
In addition to the famous decision in the case of *Roe v. Wade*, other examples include the decisions in the cases of *Eisenstadt v. Baird*, 405 US 438, 453 (1973); *Clevelet Board of Education v. Le Eleur*, 414 US 632, 639 (1974); and *Planned Parenthood v. Casey*, 505 US 833, 851 (1992).

to the philosophical and legal debates that take place at specialist congresses and other meetings.

The most radical and controversial argument against cloning concerns the embryo's "right to life", and here the debate on therapeutic cloning clearly overlaps with the abortion debate. It is an argument that tends to appear in preparatory studies and explanatory memoranda rather than laws themselves. As already noted, however, the Irish Constitution, in particular, guarantees to defend "the right to life of the unborn".

For opponents of anti-cloning laws, some of whom are the same people who support "free choice" in the abortion debate, embryos are devoid of all characteristics of human life and have no status as human beings. At least, this is what they maintain in respect of the aggregation of undifferentiated cells in the first few days after conception, which is when any interventions for therapeutic cloning take place.

It has also been claimed that the embryo's right to life would seem to be contested, if not denied, in international human rights treaties ratified by the countries concerned. The second part of this chapter looks in more detail at this subject.

One apparently similar but in fact quite separate argument is the claim that embyros must be protected because they carry within them the potential for human life. In the course of the preparatory work to draw up the Council of Europe's Additional Protocol, France stated that "It may plausibly be argued that the ban on infringements of human dignity proclaimed by French legislation is applicable even if the 'person' concerned, at the time of the cloning operation, is a potential person ..."[1]

This statement highlights the principle usually upheld in laws restricting cloning, namely the principle of respect for human dignity. It is used in Switzerland's new Constitution, for example, to justify the ban on "all forms of cloning". In France, this principle, in particular as laid down in Article 16 of the Civil Code, would appear to have acquired constitutional status following the decision handed down by the Constitutional Council on 27 July 1994 on the bioethics laws. In its opinion of 22 April 1997, the French Advisory Committee on Ethics held that, irrespective of the end purpose and technique used, all

1.
Doc. CDBI/Inf(98)8, *op.cit.*, p. 226.

forms of cloning that consist in introducing asexual reproduction into the human species radically undermine human
dignity.

Some lawyers have criticised clauses or statements such as
these for being dogmatic and obscure.[1] Precisely in what way,
they ask, would the mere fact of cloning undermine the repu-
tation and self-esteem of the people concerned?

Supporters of the "human dignity" clauses fear the "instrumentalisation"[2] of human beings. Others, while not denying that
cloning may give rise to reprehensible practices similar to
slavery – for the sole purpose, for example, of setting up a source
of transplant organs – have pointed out that there is always a
danger that biological advances may be abused. So as not to
"throw out the baby with the bathwater", it was important to
deal separately with each advance and the risks involved.

A number of laws, including once again the Swiss Constitution,
highlight risks with respect to "personality" or "individuality". It
has been pointed out, however, that even in the case of identical twins brought up in the same family and attending the
same schools their personalities are never absolutely identical.
Like the biological contribution of the mother, differences
between clones and their creator, in terms of their environment and experiences, would prevent the development of
mere replicas.

Recourse in law to a similar concept, namely that of personal
"integrity", has also been criticised. Several commentators have
said it is not clear in what way or to what extent any of the
three parties involved in the cloning operation might be
affected, physically or mentally, by the operation.[3]

Various laws restricting cloning have mentioned the risks of
improper selection and "eugenism". Opponents have acknowledged that such a danger exists but without going so far as to
admit that it is a risk specific to cloning or more of a threat
with cloning than with other reproductive techniques. In particular, pre-implantation diagnosis could carry the same, if not
a greater, risk of eugenism in a society where attitudes towards
abortion are increasingly permissive. Nonetheless, the public
chooses to focus most of its attention on cloning.

1.
See the articles by
John A. Robertson,
"Reproductive Liberty
and the Right to
Clone Human Beings"
and John Harris,
"Clones, Genes and
Reproductive Auton-
omy" in: *Ann NY.
Acad. Sci.*, September
2000.

2.
See the Group of
Advisers on the Ethical
Implications of
Biotechnology, report
submitted to the
European Commis-
sion, Opinion No. 9,
28 May 1997.

3.
See for example:
"Human somatic cell
nuclear transfer",
report by the Ethics
Committee of the
American Society for
Reproductive Medi-
cine of 7 August 2000
in *Fertility and Steril-
ity*,. 15, No. 5,
November 2000,
pp. 274-275.

Genetic recombination:
this occurs during meiosis and generates further variation between gametes. Homologous chromosomes exchange parts and thus produce new combinations of genetic material.

The risks of serious health impairments and deficient clone development have been highlighted both in legislative debates and at science meetings.

Lastly, it has been claimed in support of laws restricting cloning that natural genetic recombination* would offer human beings a higher degree of freedom than some predetermined genetic make-up and that it would therefore be in all our interests to keep the essentially random nature of our genes.[1] It has been pointed out, however, that this risk ought to be minimal, insofar as, even with the benefit of technical advances, cloning will most likely remain an exceptional method of human reproduction.[2]

This brief analysis shows that there is little or no mention in anti-cloning laws and preparatory studies of certain situations, such as sterility in the case of couples in stable relationships and wishing to start a family, even though some commentators consider them justification for reproductive cloning.

It is also striking that these laws should reject cloning outright, even when it is carried out for therapeutic purposes. The benefits of these techniques for the advancement of human health are not denied but they are banned on the basis of two alternative arguments: the "right to life of the unborn" (see above) and the danger that therapeutic cloning will lead to reproductive cloning, since up to a point the biological process is the same.

Acceptance of cloning for therapeutic or research purposes

1.
See for example, Council of Europe, Explanatory Report on the Additional Protocol on the Prohibition of Cloning Human Beings, 25 July 2001, paragraph 3.

2.
Report by the Ethics Committee of the American Society for Reproductive Medicine, *op. cit.*, and aforementioned authorities.

In some countries, cloning is prohibited if its purpose is to create a human being but allowed, subject to certain conditions and restrictions, if it is carried out for research or curative purposes.

In the United Kingdom, reproductive cloning is prohibited by the Human Reproductive Cloning Act 2001 which makes it an offence for a person to place into a woman a human embryo which has been created otherwise than by fertilisation.

With respect to therapeutic cloning, a new statutory instrument entitled "The Human Fertilisation and Embryology (Research Purposes) Regulations 2001" could allow such a procedure to take place. However, in this last instrument a licence would need to be given before any research is undertaken by the Human Fertilisation and Embryology Authority, a public body set up in 1990, consisting of scientists and experts in the field of ethics.

In the Netherlands, there is a similar proposal for a law banning reproductive cloning but allowing *in vitro* research on embryos under certain conditions, notably if such research is likely to have significant benefits for medical science.

Only very recently French legislation has shown signs of moving in a similar direction.

In China, according to unofficial information, it would seem that in December 2000 a scientific advisory committee on the human genome proposed banning reproductive cloning while allowing research conducted with a view to producing human organs for medical purposes.[1]

Laws of this kind recognise that some embryonic research and therapeutic cloning can be of great value for promoting the right of all individuals to enjoy the best possible health. Priority is given, in our view quite rightly, to this basic human right over the right to "life" and "dignity" of the unborn relied on in the laws identified in the previous section.

However, aspects relating to other human rights cited as justification for reproductive cloning, such as the right to raise a family, are still considered in these legal systems to be secondary compared with the threat of instrumentalisation and the other arguments examined on which the ban on reproductive cloning is based.

On the whole, these recent legislative developments would appear to be moving in the right direction. They make necessary distinctions in favour of the right to health in an area of the law where it seemed that the tendency was to reject everything *en masse*.

Moratorium approach

Several experts have advocated the introduction of a moratorium during which time certain interventions on embryos

1.
Nature Medicine,
No. 5, Vol. 7, May
2001.

would be banned in the interests of a more thorough and impartial analysis of all the scientific, ethical, social and legal issues relating to cloning.[1]

A five-year moratorium on reproductive cloning was announced for example in Israel[2] in an act of 29 December 1998. An advisory committee of experts was instructed to monitor scientific progress in this field and to submit an annual report and recommendations to the Minister for Health. On the basis of one of these recommendations, and providing he or she considered there was no violation of human dignity, the Minister for Health could authorise certain kinds of reproductive cloning subject to certain reservations and control.

In the Russian Federation, a bill for the introduction of a five-year moratorium on cloning of human beings was brought before the Duma (lower house of parliament) in August 2001.

This approach has the considerable advantage of allowing serious studies and debate gradually to take the place of emotional reactions. Providing its purpose is well defined and an effective means of control are in place – which implies the need for political resolve as well as scientific and budgetary resources – a moratorium can be the appropriate response in the present circumstances.

It is by no means a panacea, however. Animal testing can never provide any absolute guarantee that the same techniques can be used on humans.

It is regrettable that so few laws have established a moratorium.

1.
See the report by the Ethics Committee of the American Society for Reproductive Medicine and aforementioned authorities, *op. cit.*, p. 875.

2.
Sefer Hacherkkim, 7 January 1999, No. 1697, p. 47.

Cloning and international human rights law

Treaties

The Council of Europe's Convention and Additional Protocol

The Additional Protocol to the Convention for the Protection of Human Rights and Dignity of the Human Being with regard to the Application of Biology and Medicine on the Prohibition of Cloning Human Beings, entered into force on 1 March 2001.

On 8 April 2002 the contracting parties who had ratified it were Cyprus, Czech Republic, Estonia, Georgia, Greece, Hungary, Portugal, Romania, Slovakia, Slovenia and Spain. The potentially far-reaching geographical scope of this instrument must be stressed: it is open to all contracting parties to the 1997 parent Convention on Human Rights and Biomedicine, including those non-member states which participated in its elaboration,[1] and also to other non-member states by special decision of the Committee of Ministers.[2]

The first article of the Additional Protocol,[3] which is the only binding international instrument on the subject, prohibits "any intervention seeking to create a human being genetically identical to another human being, whether living or dead". The term "genetically identical" is defined as meaning "a human being sharing with another the same nuclear gene set". No derogations are permitted.

The Additional Protocol is subject to the same international implementation procedure as the convention, that is, a system of government reports examined by a committee of experts. The European Court of Human Rights may give advisory opinions on interpretation of the treaty when so requested by a contracting party or by the committee of experts, subject to certain conditions. The Court is not empowered, however, to receive applications concerning alleged violations of the Additional Protocol or the convention.

Other relevant human rights treaties

The Council of Europe's Additional Protocol to the Convention on Human Rights and Biomedicine is at present the only intergovernmental treaty which expressly addresses the issue of cloning.

It should be noted that at least one European Union directive also mentions it, namely the 1998 Directive on the legal protection of biotechnological inventions,[4] which excludes "processes for cloning human beings" from patentability on the grounds that such cloning "offends against public order and morality".

1.
Art. 33.1 of the convention.

2.
Art. 34 of the convention.

3.
European Treaties Series No. 168. See Appendix III.

4.
Directive 98/44/EC of 6 July 1998, *O.J. of the European Communities* No. L213, 30/07/98, pp. 13 to 21.

It is conceivable that the Court of Justice of the European Communities or other EU bodies might in future produce binding texts of a wider scope on cloning and interventions on the embryo, relying for this purpose on the Maastricht and Amsterdam treaties and the case-law of the European Court of Human Rights in Strasbourg. Article 3.2 of the EU "Charter of Fundamental Rights" adopted at the Nice Summit in 2000 already prohibits "the reproductive cloning of human beings", but the Charter is merely a declaratory text.

In addition to this specific EU dimension, numerous treaties pre-dating the Protocol could possibly be taken into consideration. Admittedly, they do not mention cloning as such, for the simple reason that they were signed at a time when such things often seemed to belong to the realm of science fiction. Also, the context in which these instruments were framed could lead to their application to cloning being considered inappropriate under Article 31 of the UN Vienna Convention on the Law of Treaties. Nevertheless, the preparatory documents show that their authors often intended them to cover all future situations, including scientific advances in areas where rapid progress could be foreseen. For this reason I think these general treaties on human rights deserve a brief mention here.

It goes without saying that these general conventions, to the extent that they could be interpreted as accepting reproductive cloning, would be inapplicable for the parties to both treaties because they pre-date the Council of Europe's Additional Protocol.[1] However, their relevance would still be worth considering in respect of states that are not parties to the Additional Protocol, and even for those that have ratified it, with regard to other forms of cloning.[2]

In support of the prohibition on cloning, mention has been made of the preamble to the UN Convention on the Rights of the Child, the ninth paragraph of which states that the child "needs special safeguards and care, including appropriate legal protection, before as well as after birth".

Some states see this clause as enshrining the right to life from the moment of conception, which would exclude all cloning. This is a debatable interpretation, however, as the preamble

1.
Article 30 of the UN Vienna Convention on the Law of Treaties.
See: www.un.org/law/ilc/texts/treaties.htm

was adopted subject to the formal rejection of amendments along those lines of Article 1 of the convention, which defines the "child".[1]

The right to life, "in general from the moment of conception", is proclaimed in Article 4.1 of the 1969 American Convention on Human Rights. However, the Inter-American Commission on Human Rights appears, in practice, to have qualified the scope of this provision.[2]

By contrast, as we shall see a little later, other conventions have been based on explicit rejection of the right to life from the moment of conception.

Could slavery, defined in the 1919 and 1926 conventions as "the state or the position of the person, over whom the attributes of the right of ownership or some of them are exercised"[3] be likened legally to the status of a clone, physically or psychologically conditioned to serve exclusively as an instrument for the personal designs of another? This would be a bonus for those who oppose cloning, as the prohibition of slavery has been raised to *jus cogens* status; that is, a standard from which no derogation is possible, by the United Nations International Law Commission.

However, such concerns are alien to the context and drafting history of the anti-slavery treaties and the legitimacy of such an interpretation, likening cloning to wrongful treatment of human beings, has yet to be proved.

Another argument relied on by opponents of cloning is the prohibition of "torture and cruel, inhuman or degrading treatment or punishment", to quote the wording used in the Universal Declaration of Human Rights and reiterated in Article 7 of the UN Covenant on Civil and Political Rights, Article 3 of the Council of Europe's European Convention on Human Rights and numerous other instruments.

Article 7 of the covenant also prohibits medical experimentation without the patient's consent. It has been argued that bringing a clone into the world without its consent, and the serious health problems it would suffer, would violate these rules.[4] It is possible that the European Court of Human Rights

1.
See report of Working Group on the Convention of the United Nations Commission on Human Rights, 1989, United Nations doc. E/CN4/1989/48, paragraphs 32-47 and 75-85 and appended Opinion of the Legal Adviser to the United Nations.

2.
See Res. No. 23/81, case 2141 (USA), Inter-American Convention on Human Rights, 1980-81 annual report, doc. OEA/Ser. L/V/II.54, doc. 9 rev. 1, 16 October 1981.

3.
Article 1 of the 1926 Slavery Convention, which entered into force on 9 March 1927, in accordance with Article 12.

4.
See, for example, the Report of the Ethics Committee of the American Society for Reproductive Medicine, *op. cit.*, p. 874.

will be required to rule on these issues. The health problems affecting clones may, however, be reduced or eliminated in the future.

The right to enjoyment of the "highest attainable standard of health", enshrined in Article 12 of the UN Covenant on Economic, Social and Cultural Rights and in other treaties (1966), is another argument that could be used by opponents of cloning, but not in a decisive manner as the article is ambivalent.

Those in favour of a more flexible attitude towards cloning may turn to a number of texts, many of which show the same ambiguity.

Neither the UN Covenant on Civil and Political Rights nor the European Convention on Human Rights recognises the right to life "from the moment of conception". Amendments along those lines to Article 6 of the covenant were rejected by formal vote.[1]

Can the "human freedoms" enshrined in the covenant, the European Convention on Human Rights and all human rights treaties be considered, in general, to include the right of any person to clone him or herself? It should be noted that a related reference to personal "integrity" proposed for the UN Universal Declaration of Human Rights was omitted from all subsequent versions.[2] It should also be noted that, in the covenant, the principle of freedom is not in itself subject to any restrictions.[3]

However, the principle of freedom is so closely linked with problems of detention in these treaties that taking it out of this context might be considered inappropriate.

The right to privacy, which is also enshrined in all human rights instruments, has often been relied upon in proceedings before the European Court of Human Rights to justify freedom in respect of human reproduction, particularly in cases concerning abortion and the status of transsexuals. The Court – and the Commission in its day – acknowledged this freedom, while at the same time stressing the broad margin of appreciation enjoyed by states with regard to the restrictions provided for in the treaties themselves in the interests of public order, public morality, etc.[4]

1. Amendment tabled at the General Assembly by Belgium, Brazil, Morocco, Mexico and El Salvador and rejected by 20 votes in favour, 31 against and 17 abstentions, A/C.3/SR.820.

2. E/CN4/AC.2/SR.3, 6 December 1947.

3. Under the derogation clause (Art. 4), however, it may be suspended temporarily "in time of public emergency which threatens the life of the nation".

4. See, for example, the then Commission's decision in the case of *Brüggemann and Scheuten v. the Federal Republic of Germany*, 1977, 3 *European Human Rights Reports*, 244.

The right "to marry and to found a family" (Article 23 of the UN Covenant on Civil and Political Rights, Article 12 of the European Convention on Human Rights) and the separate right to protection of one's "family life" (Article 17 of the UN Covenant on Civil and Political Rights, Article 8 of the European Convention on Human Rights), which are also universally recognised, have been the subject of many decisions by the European Court of Human Rights. For a long time, Article 12 of the European Convention on Human Rights was considered to refer solely to coital and sexual reproduction, but changes in the Court's case-law seem likely, provided there is stability in the couple's relationship. The right to respect for family life has been the subject of some highly nuanced decisions. However, although the Court has independent status, can it really be expected to ignore the prohibitions laid down in the Convention on Human Rights and Biomedicine and its Additional Protocol, which are also Council of Europe instruments?

The right to enjoy the highest attainable standard of health is one of the major arguments used to justify cloning for medical and therapeutic purposes. Article 12 of the UN Covenant on Economic, Social and Cultural Rights may seem to have been drafted in terms general enough to include this type of cloning. The preamble and explanatory report to the Additional Protocol to the Convention on Human Rights and Biomedicine seem to confirm its lawfulness under international law.

Article 16 of the Covenant on Economic, Social and Cultural Rights acknowledges the right "to enjoy the benefits of scientific progress and its applications". It further stipulates, in deliberately strong terms that suggest immediate application, that "the States Parties undertake to respect the freedom indispensable for scientific research".[1]

The freedom to conduct scientific research may be subject "only to such limitations as are determined by law only in so far as this may be compatible with the nature of these rights and solely for the purpose of promoting the general welfare in a democratic society" (Article 4). Severe restrictions on research into cloning, including cloning for therapeutic purposes, which

1.
See UN document A/2929 TP, Comments on the draft Covenants, 1955, p. 155, paragraph 55.

seem to be enforced in certain countries (see above), and punishment thereof as a criminal offence, would be difficult to reconcile with Article 4 of the covenant.

The fact remains that this liberal clause concerning research would be subject, for those states which are parties to all three instruments, to the tighter restrictions and the prohibitions laid down in the 1997 Convention on Human Rights and Biomedicine (Articles 15 to 18) and in the 1998 Additional Protocol.

Finally, one could – and in my opinion should – look very closely at the applicability in this context of the principles of equality of opportunity, non-discrimination and equality before the law, all of which are included in one form or another in all the human rights instruments. Article 26 of the Covenant on Civil and Political Rights, on equality before the law, is particularly useful, as it may even cover rights not included in this covenant, such as the right to health and freedom to conduct research, which are recognised by the Covenant on Economic, Social and Cultural Rights.

For example, one should consider the applicability of these principles in respect of the refusal to allow sick people to enjoy the benefits of cloning for therapeutic purposes even where this is the only treatment possible and their only hope of recovery, when others, suffering from the same diseases but amenable to other types of treatment, survive.[1]

1.
The case-law of the UN Human Rights Committee, which is responsible for monitoring the implementation of the Covenant on Civil and Political Rights, tends, it is true, to exclude Article 26 when the situations are not identical. There must be limits to this reasoning, however, when people's health and lives are at stake.

Customary international law?

In both domestic law and international treaties there has been a tendency since 1997 to prohibit reproductive cloning. This is not the case, however, in all the countries which have passed legislation on the question. There is also a wide variety of legislative approaches to cloning for therapeutic purposes revolving around often ill-defined concepts. Finally, it should not be forgotten that in a very large number of countries, particularly in Africa, Asia and, especially, central and eastern Europe, there is still a legal vacuum.

The directly or indirectly relevant declaratory texts of inter-governmental organisations are themselves very imprecise and varied in their approach.

On the one hand, there are those that lean generally towards the prohibition of cloning, such as the Resolutions of the WHO General Assembly since 1997 and the Unesco Declaration on the Human Genome and Human Rights. On the other hand, the declarations and action programmes of the major UN world conferences on population (Bucharest and Mexico), gender equality (Nairobi and Beijing) and other subjects increasingly emphasise freedom of choice where procreation is concerned. The word "cloning" is not actually mentioned, but some passages could be interpreted as not being opposed in principle to cloning, for therapeutic purposes at least.

Clearly, therefore, the conditions are not right for the development of a body of customary international law on cloning or genetic engineering in general.

Spurred on by media coverage and emotional public reactions, the standard-setting trend with regard to cloning issues has developed very quickly, no doubt too quickly in some countries. For once, the law may seem to be moving faster on an issue than scientific experimentation.

This phenomenon may seem regrettable in various respects. Once enacted, laws tend to fix attitudes and make them more rigid. Reforms and adjustments become more difficult to implement.

Before continuing to lay down rules, law-makers and politicians should pay greater heed to the voice of the scientific community, which is calling widely for more reflection on the physical, psychological, social and ethical aspects of cloning.

Nobody denies that cloning techniques are imperfect at present, and nobody denies the health risks to clones or the danger that cloning will be abused and exploited for criminal, political or commercial purposes. However, should we not distinguish between cloning itself and its exploitation in violation of human rights? Comparable risks of abuse exist in connection with other advances in biology, against which we try to provide safeguards, but without condemning them outright.

A moratorium would make sense, to enable study groups and think tanks composed of scientists, doctors, legal experts, ethicists and parliamentarians to consider cloning from every angle and put together an action plan. This would lay the foundations for a rational policy on cloning which would also be consistent with human rights.

Bibliography

American Society for Reproductive Medicine, "Human Somatic Cell Nuclear Transfer", in *Fertility and Sterility*, vol. 74, No. 5, novembre 2000.

Council of Europe, *Assistance médicale à la procréation et protection de l'embryon humain, étude comparative dans 39 pays; clonage, étude comparative dans 44 pays,* Doc. CDBI/INF (98) 8, 2 juin 1998.

Council of Europe, *Protocole additionnel à la Convention sur les droits de l'homme et la biomédecine, texte et rapport explicatif portant interdiction du clonage d'êtres humains,* Série des traités européens, No. 168, 1998.

Harris, J., "Clones, Genes and Reproductive Autonomy, the Ethics of Human Cloning", in *Annals of the New York Academy of Sciences,* September 2000.

Kahn, Axel, "Aspects éthiques du clonage humain à finalité thérapeutique et de l'utilisation des cellules souches embryonnaires", in *Bulletin de l'Académie nationale de médecine,* No. 6, 2000, p. 184.

Mattei, Jean-François, "La question éthique de l'embryon", in *Bulletin de l'Académie nationale de médecine,* No. 6, 2000, p.1139.

United Nations Organisation, *Commentaire des projets de pactes relatifs aux droits de l'homme par le Secrétaire Général,* Doc. A/2929, 1955.

United Nations Organisation, *Normes des Nations Unies concernant les rapports entre les droits de l'homme et la population,* Doc. E/Conf. 60/SYM. IV/3, 1974, et document de mise à jour, préparés par Maxime Tardu, expert-consultant, 1988.

United Nations Organisation, Commission des droits de l'homme, *Recueil d'instruments internationaux,* deux volumes, 1993.

World Health Organisation, Health Legislation Unit, *Informal Listing of Laws, Regulations, Guidelines on Cloning,* 12 May 1998.

World Health Organisation, *Recueil international de législation sanitaire*, 1998-2000.

World Health Organisation, Résolution OMS 50.37 du 14 mai 1997 sur le clonage dans la reproduction humaine, in *Manuel des résolutions*, vol. III (3rd edition).

Petitti, L.-E., Decaux, E. et Imbert, J., *La Convention européenne des Droits de l'Homme*, sous articles 8 et 12, Paris, Economica, 1995.

Robertson, J.A., "Reproductive Liberty and the Right to Clone Human Beings", in *Annals of the New York Academy of Sciences*, September 2000.

Unesco, *Déclaration universelle sur le génome humain et les droits de l'homme*, 11 November 1997.

The Council of Europe's position

Reproductive cloning

On 12 January 1998 the Council of Europe opened for signature an Additional Protocol to the Convention on Human Rights and Biomedicine,[1] on the Prohibition of Cloning Human Beings. Article 1 of this protocol stipulates that: "Any intervention seeking to create a human being genetically identical to another human being, whether living or dead, is prohibited". Paragraph 2 of the same article states that "the term human being "genetically identical" to another human being means a human being sharing with another the same nuclear gene set". The creation of a genetically identical human being is prohibited, irrespective of the technique used (embryo splitting or nuclear transfer). It is, in fact, not the technique but the ultimate aim that is prohibited.

In other words, since 1998 the Council of Europe has prohibited any attempt at reproductive cloning aimed at producing a cloned human being. This Additional Protocol came into force on 1 March 2001 and has been signed by 29 out of the 44 member states of the Council of Europe. The Additional Protocol has now been ratified by, and has entered into force in, eleven countries[2] but other countries have signified their intention to ratify it as soon as possible.

When a number of scientists announced their intention to clone human beings for reproductive purposes, governments realised that there was an urgent need to prohibit this practice as it would inevitably be incompatible with a number of fundamental principles such as human dignity.

So-called "therapeutic cloning"

Although there is general agreement on condemning reproductive cloning, not all countries agree on the need to prohibit so-called "therapeutic cloning", that is, the creation of embryos through cloning as a source of new cells for use in medical treatment.

1.
Full title of the parent convention: The Convention for the Protection of Human Rights and Dignity of the Human Being with regard to the Application of Biology and Medicine.
See: http://book.coe.fr/conv/en/ui/ctrl/menu-en.htm. See also Appendix III.

2.
Cyprus, Czech Republic, Estonia, Georgia, Greece, Hungary, Portugal, Romania, Slovakia, Slovenia, Spain.

In answer to the question whether this Additional Protocol permits the prohibition of therapeutic cloning, it should be pointed out that the wording of the aforementioned Article 1, uses the term "human being". Therapeutic cloning is therefore prohibited in all countries that consider an embryo a human being. It should, nevertheless, be noted that not all Council of Europe member states agree on this, for example, when signing the Additional Protocol, the Netherlands deposited a declaration, dated 29 April 1998, with the Secretary General, stating that it interpreted the term "human being" as referring exclusively to a human individual, that is, a human being who has been born. Consequently, the Additional Protocol prohibiting the cloning of human beings does not necessarily prohibit therapeutic cloning.

Nevertheless, for the time being and for some years to come, human cloning will require a considerable amount of research and is in fact, in itself, a form of research. "The creation of embryos for therapeutic purposes" can therefore be equated with "the creation of embryos for research purposes", which is prohibited by the Convention on Human Rights and Biomedicine (done at Oviedo, 4 April 1997). Article 18.2 of the convention, which came into force on 1 December 1999, stipulates that "the creation of human embryos for research purposes is prohibited." It should also be mentioned, however, that any country signing the convention can enter a reservation to Article 18.2 if they have (as the UK has) a law allowing the creation of human embryos for research purposes.

It is, nevertheless, impossible to ignore the fact that a number of private companies, not only in the United States but also in certain European countries, are carrying out active research into the possibilities of human cloning. Even if this is a distant prospect, which would take cloning for therapeutic purposes beyond a purely research framework, the Council of Europe must look to the future by taking a clear-cut position on the creation and subsequent use of embryos for therapeutic purposes, particularly in the context of its draft Protocol on the Protection of the Human Embryo and Foetus. It will need to be borne in mind that there are other ways of obtaining cell lines.

There are, in fact, good reasons for encouraging the use of adult stem cells, and research – in particular public research – should make greater use of this method.

Towards a worldwide ban on human cloning?

by André Albert

The possibility that cloning techniques may be applied to human reproduction has sparked a debate which differs in three respects from the earlier debates on many more "conventional" ethical issues.

First, unlike the controversy roused by the use of certain medically assisted techniques of reproduction, this one has started before it is even sure that the nuclear transfer technique used to clone Dolly the sheep will actually work for human beings.

Second, it has resulted in immediate condemnation of any such project. This condemnation may not be universal, but it does transcend most of the disagreements which normally arise between communities, cultures and religions on other bioethical problems.

Finally, it has very quickly been made plain that regional banning is no answer, and that this is a problem which must be solved at world level.

There is no point in going back here over the efforts which various international organisations have been making since 1997 to secure a ban on any attempt to produce a human child by cloning.[1] Such a ban has now been solemnly incorporated in a number of international instruments, including Unesco's Universal Declaration on the Human Genome and Human Rights, the Council of Europe's Additional Protocol to the Convention on Human Rights and Biomedicine, prohibiting the cloning of human beings,[2] and the European Union's Charter of Fundamental Rights.

Important as these texts are, not one of them meets the need for a ban which is both universal and enforceable.

In the meantime – and the start of 2002 brought a flood of such announcements – various biologists, some of them backed by sectarian groups, have been telling the media that

1.
This is what the commoner term "reproductive cloning" signifies. This term is itself ambiguous, however, since cloning is always "reproductive" – of organs and organisms, or cells, to use the definition made by Henri Atlan in his contribution to Le clonage humain, Paris, Le Seuil, 1999.

2.
Unlike the Unesco Declaration on the Human Genome and Human Rights and the European Union's Charter of Fundamental Rights, the Additional Protocol to the Council of Europe's Convention on Human Rights and Biomedicine does not limit its ban to "reproductive" cloning. However, this does not necessarily mean that it prohibits "therapeutic" cloning, – at least if this ambiguous term – means the creation of embryonic stem cell lines by transferring an adult cell nucleus to an enucleated oocyte.

they mean to be the first to produce a child by cloning. They talk about laboratories in secret locations, where no relevant laws apply – all of which highlights the present lack of instruments to deter and punish worldwide.

French-German initiative

It was in this context that France and Germany jointly approached the UN in autumn 2001, asking it to start work on a convention to ban cloning for human reproductive purposes, backed by effective sanctions and providing for international co-operation between police forces and courts.

A special committee of the UN's 6th Committee,[1] on which some 120 of the UN countries were represented, held a first exploratory meeting on this question from 25 February to 1 March 2002. It will be meeting again from 23 to 27 September 2002, to finalise a negotiation mandate for submission to the UN General Assembly by the end of 2002.

Apart from a few delegations which reserved their positions or wanted a moratorium, national reactions at the first meeting reflected a clear division between states which favoured variants on the Franco-German position (mandate restricted to "reproductive" cloning) and states which sought a total ban on nuclear transfer, for example, the United States, the Vatican, as well as, with certain qualifications, two other European states.

There is a clear difference between the reasons for these two positions. Those who want a total ban rely on fundamental arguments, insisting that the human embryo must be protected, and that a ban on "reproductive" cloning only may not work, since nuclear transfer will still be authorised for certain purposes. The Franco-German proposal, on the other hand – without prejudging a basic position on "therapeutic" cloning – is essentially strategic in its aims: only a ban initially limited to "reproductive" cloning is sure of rapidly commanding a consensus broad enough to parry a situation in which law and practice are racing neck-and-neck, by imposing an unconditional prohibition, and not just some kind of moratorium.

1.
It should be remembered that the 6th Committee, the UN's legal committee, is special in taking its decisions by consensus, rather than a vote.

Current positions

In fact, reactions at the meeting in February 2002 give an idea of some of the areas of disagreement, and of the way in which an instrument which set out to ban both "reproductive" and "therapeutic" cloning might fragment the original problem.

Positions are actually far more varied on "therapeutic" than on "reproductive" cloning: some countries argue for approval, others want a ban – more or less provisional, more or less radical – and others again are in the middle, calling for various forms and levels of regulation.

National positions on "therapeutic" cloning are also more complex: some of the questions raised are more general (for exemple safeguards for embryos *in vitro*, regardless of their origin), others go beyond bioethics (acceptable restrictions on freedom of research, application of the precautionary principle to cloning, or even North-South relations, [1] etc.).

This shows that, by setting out to cover all the aims of cloning in a single instrument, one runs the risk of failing to achieve sufficient consensus and a clear organising principle, and so wandering off-course and ending up with a simple moratorium. This, clearly, is not what the Franco-German initiative is aiming at, since a moratorium on "reproductive" cloning would certainly be taken as implying its acceptance in the long term.

Of course, the convention which is now being considered could incorporate preventive measures aimed at "reproductive" cloning, and suggest that practices which could be diverted in this direction, such as research on embryo stem cells, be strictly regulated.

Nonetheless, to avoid fragmenting discussions and to keep the issues clear, it would seem essential, not just to focus on prohibition of one thing, but also to base that prohibition on a recognised, universal value.

A challenge for civilisation

Occasionally, the force of the various arguments in favour of an absolute, permanent ban on "reproductive" cloning, put

1.
The North/South element in the debate chiefly concerns the need to protect potential oocyte donors in the developing countries, and the danger that the industrial countries may monopolise the benefits of new cell therapies based on nuclear transfer – that is, if such therapies ever came to be developed.

forward by the experts at the meeting in February 2002, almost seemed to be weakened by their sheer multiplicity.[1]

At the risk of stating the obvious, it is doubtful whether we can secure a universal ban on "reproductive" cloning quickly, without first deciding what basis to give it.

The experts referred to a whole series of potential dangers, and indeed too many. Taken together, these dangers were alarming - particularly in terms of public health and harmonious development of the child. Taking them singly, however, one had the impression that, if better answers could only be found, various exceptions to the total ban on "reproductive" cloning might yet prove acceptable.

The trouble with this approach is that it fails to make an absolute, universal ban seem necessary.

When the gametes of a man and woman meet in procreation, two sets of inherited genes recombine to produce a child whose characteristics cannot be foreseen.

It is precisely this random recombination – possible in sexual reproduction – which "reproductive" cloning interferes with. Of course, "random" sexual reproduction is not peculiar to humans, but casting doubt on it in humans opens the way to radical revision of their symbolic status, and the full consequences of this cannot be predicted, or even imagined.

In practice, cloned children may be genetically predetermined by the people who decide to produce them and who are not, strictly speaking, their parents, or may find their future firmly mapped out in the minds of those around them. The unpredictable traits conditioned by the individual genome will be lacking, and nothing will be left for them to discover in themselves, and make sense of freely in their own projects, regardless of what other people or the community think or know about them.

It is perfectly possible, of course, that a person who has to face the fact that (s)he is another person's clone may come through that personal test unscathed, without loss of individuality or freedom. But an essential element in other people's *recognition* of that individuality and freedom will still be missing. Because

1.
The independent experts' main arguments were : the medical risks attendant today on the possible use of nuclear transfer techniques for purposes of human reproduction, the threat to the rights of oocyte donors, and the danger that the cloned child's genetic identity with the adult from whom he/she has been cloned – in some ways, a faithful mirror of his/her somatic and medical future – may prove a source of serious psychological disturbance. However, none of these arguments, taken on its own, seemed an adequate basis for an absolute, permanent ban.

2.
People who cite monozygotic twins (see def. p. 41) as a counter-argument need to be reminded that the genetic status of such twins is not deliberately chosen. It should also be emphasised that it is elimination of the random element in human procreation, and not genetic identity as such, which is the decisive reason for banning "reproductive" cloning.

of his/her special status,[1] such a person will inevitably be exposed to far greater risks of discrimination, instrumentalisation and even servitude in his/her dealings with individuals, groups and the community, than people whose birth allows a biological principle of uncertainty and unpredictability to operate freely.

No special circumstances, no back-up measures, and no individual or institutional safeguards – parental, familial, medical, sociological, legal or political – can justify a situation in which this status is forced on certain people, who are then saddled with a serious in-built handicap in their efforts to achieve individual autonomy.

Of course, the cogency of this ground for prohibition has frequently been recognised, and some of the legal texts on human cloning have, very rightly, treated it as essential.[1] At the same time, neither the things which differentiate this argument from those used in the debate on "therapeutic" cloning,[2] nor the fact that it admits no exceptions to the principle of prohibition, nor – and this is the most important point – its highlighting of the fact that banning "reproductive" cloning is a matter of upholding civilised values, seems to have been brought out sufficiently in public discussion of these issues.

1.
This is particularly true of the Additional Protocol to the Council of Europe's Convention on Human Rights and Biomedicine, as paragraph 3 of the explanatory report makes clear.

2.
The societal issues at stake in protection of embryos *in vitro*, and specifically the general principle of prohibiting the production of embryos for research purposes, can scarcely be ranked with those at stake in the banning of "reproductive" cloning.

Conclusion

by Dr Anne McLaren

"To clone or not to clone, that is the question"

Could we?

For some animal species, we know that we can. Sheep, goats, cows, pigs, cats and mice have all been cloned by nuclear transfer, but the efficiency is still very low. For other species, it is possible that somatic cell nuclear transfer cloning will never be achieved. In dogs and monkeys, serious attempts have been made, but no liveborn young have been reported.

For humans, we do not know. Few attempts have been made: human eggs are not easily obtained, and in many countries such research is against the law. The only paper to be published in a scientific journal reported that the reconstructed eggs either failed entirely to divide, or divided just a couple of times and then stopped. This degree of development provides no evidence that the transferred somatic cell nucleus was to any extent reactivated or reprogrammed by the egg cytoplasm.

Should we?

For non-human animals, the ethical considerations are those that pertain to all animal experimentation, in particular whether the likely benefits to humans outweigh any possible suffering to the animals. Strict regulation and ethical committee review are desirable, but there is no more reason to expect somatic cell nuclear transfer to cause more suffering than any other form of research on animals. The deaths of cloned embryos occur mainly before implantation or during gestation. If abnormalities appear after birth that are likely to cause suffering, the animals should be humanely killed.

The benefits to be derived from research on cloning animals are enormous. In mice and perhaps other laboratory animals such research could increase our understanding of ageing, cancer, the control of gene expression and the origin of congenital abnormalities. Perhaps if we understood how a highly specialised somatic nucleus can be reprogrammed by egg

Transgenic animals:
foreign genes are introduced into an organism by injecting the genes into newly fertilised eggs. Some of the animals that develop from the injected egg (transgenic animals) will carry the foreign genes in their genomes and will transmit them to their progeny.

cytoplasm, we could reprogramme a patient's own somatic cells into pluripotent stem cells without any need to use human eggs or produce embryos.

In farm animals, the benefits are likely to be economic, and directed to human welfare rather than to fundamental biological understanding. Cloning of valuable animals, for example a highly prized pedigree bull, might make economic sense even today. Any improvements in the efficiency of the cloning procedure would greatly increase the economic benefit, since the value of an individual cloned animal cannot be reproduced except by cloning.

The more widespread and immediate benefit is likely to come from the enhanced possibilities of genetic manipulation that nuclear transfer offers. Transgenic animals*, with added genes producing valuable pharmaceutical and other proteins in their milk, can be produced more efficiently, using fewer animals, by nuclear transfer than by pronuclear injection. Once transferred, the new gene can be handed on by normal sexual reproduction, obviating the need for further cloning. Furthermore, genes can not only be added to cultured cells, they can also be modified or removed, and these genetic modifications can then be established in animals by nuclear transfer cloning. Thus the bovine serum albumin gene could be replaced by the gene for human serum albumin (a protein widely used after surgery and in the treatment of burns); genes coding for proteins that cause allergic reactions to cows' milk could be removed; and genes for products such as Human Factor IX and alpha-1-antitrypsin, to name just a few examples, could be added. So if we want a cheaper source (and who doesn't!) of Human Factor IX to treat patients all over the world suffering from haemophilia and other blood diseases, then at least for the moment, cloning offers the best strategy. It's good for the patients, it's good for the animals (fewer are used), and it's good for the "pharmers". One day we may harvest our Human Factor IX from plants; but not yet.

So most people would probably take the view that we should clone animals, for certain specified purposes.

For humans, the situation is very different. Even if we could carry out nuclear transfer cloning, should we? The arguments for and against somatic cell nuclear transfer cloning for stem cells have been amply set out in previous chapters. Cloning for stem cells could be an adjunct to stem cell therapy for degenerative diseases, since use of the patient's own somatic cell nuclei to make the stem cells would almost certainly avoid transplant rejection. However, the ethical problems in obtaining human eggs, and the cost of such a labour-intensive, customised form of treatment, would almost certainly render this prospect unrealistic. Other approaches to avoiding or minimising the possibility of transplant rejection are being actively researched. Where derivation of stem cell lines from human somatic cell nuclear transfer is allowed, its chief value may lie in enhancing our understanding and hence possibilities of cure of genetic diseases. A few pluripotent stem cell lines from patients would provide an indefinite amount of material for biochemical analysis and for the development of new drugs and other therapeutic approaches.

As for human reproductive cloning, cloning for babies, there is almost universal agreement that it would be criminally irresponsible to attempt such a thing at the present time. The animal experience of mortality and morbidity during gestation and at birth, and the range of defects that have occurred in cloned animals in later life, would be totally unacceptable in a human context. Claims that errors in reprogramming could be tested for seem far-fetched.

Should we ever?

Our authors come from different countries and different professions (philosophical, scientific, legal). It will become obvious to the reader that they often have different views on what is ethically acceptable and what is not. We have not attempted to harmonise their views, recognising that Europe is a multicultural community. Indeed, part of Europe's strength lies in its diversity and in the ability of its citizens to accept this diversity.

Cloning by somatic cell nuclear transfer is prohibited by law in some European countries (Germany, Austria, Norway, Iceland). Others forbid cloning for babies, but leave the door open for cloning for stem cells (UK, Netherlands, perhaps France). Others again (Italy, Greece) as yet have no laws on cloning.

The Council of Europe takes accounts of this diversity. In its Convention on Human Rights and Biomedicine, Article 18.2 prohibits the creation of human embryos for research purposes. For any member state that signs and ratifies the convention, this article would make any use of nuclear transfer cloning an offence. However, the explanatory memorandum to the convention allows any country to make a reservation in respect of Article 18.2 when ratifying the convention, provided that it already has a law in force that "is not in conformity with the provision". In the UK, for example, it is permissible to create human embryos for research purposes, if the purposes are acceptable and the project cannot be carried out in any other way, so that the UK could enter a reservation in respect of Article 18.2.

Again the Additional Protocol[1] to the convention prohibits any intervention seeking to create a human being genetically identical to another human being, that is it prohibits cloning of human beings. However, the explanatory memorandum states "it was decided to leave it to domestic law to define the scope of the expression 'human being'". Thus for some member states, ratifying the Additional Protocol would imply that any use of nuclear transfer cloning, for whatever purpose, was prohibited; while elsewhere, reproductive cloning would be prohibited but cloning for stem cells would be allowed.

We hope that this volume will stimulate our readers to think more about these profound and sensitive ethical issues. There is a need for more discussion about cloning for stem cells – whether as an adjunct to stem cell therapy, or to increase our understanding of genetic diseases. And there is still a need for discussion about "reproductive" cloning – if it ever became safe and reliable, would we condemn it in all circumstances? That's a big "if", and probably it will never come to pass; but clarity of thought in ethical matters can best be achieved by widespread discussion, before the event, rather than hasty *post hoc* judgments.

1.
See Appendix III.

Appendices

Appendix I – Some key concepts

Cloning

Which type?

There is much confusion when people see the word "clone" being used. Depending on the age of the dictionary, the definition of biological cloning can vary:

*   A group of genetically identical individuals descended from the same parent by asexual reproduction. Many plants show this by producing suckers, tubers or bulbs to colonise the area around the parent.

*   A group of genetically identical cells produced by mitotic division* from an original cell. This is where the cell creates a new set of chromosomes and splits into two "daughter" cells. This is how replacement cells are produced in the body when the old ones wear out.

*   A group of DNA molecules produced from an original length of DNA sequences produced by a bacterium or a virus using molecular biology techniques. This is what is often called molecular cloning or "DNA cloning".

*   The production of genetically identical animals by "embryo splitting". This can occur naturally at the 2-cell stage to give identical twins. In cattle, when individual cells from 4- and 8-cell embryos are implanted in different foster mothers, they can develop normally into calves and this technique has been used routinely in cattle breeding schemes for over ten years.

*   The creation of one or more genetically identical animals by transferring the nucleus of a body cell into an egg from which the nucleus has been removed. This is also known as Nuclear Transfer (NT) or Cell Nuclear Replacement (CNR) and is how Dolly was produced.

Technology

Nuclear transfer involves transferring the nucleus from a diploid cell* (containing 30 000-40 000 genes and a full set of

Mitotic division: the typical process of cell division that is preceded by chromosome replication and results in two identical "daughter" cells.

Diploid cell: a cell containing a paired set of each chromosome (46 in humans). All somatic cells are diploid.

Oocyte: the female
germ cell or egg.

paired chromosomes) to an unfertilised egg cell from which
the maternal nucleus has been removed. The technique
involves several steps (see diagram p. 179). The nucleus itself
can be transferred or the intact cell can be injected into the
oocyte*. In the latter case, the oocyte and donor cell are
normally fused and the "reconstructed embryo" activated by a
short electrical pulse. In sheep, the embryos are then cultured
for 5-6 days and those that appear to be developing normally
(usually about 10%) are implanted into foster mothers.

Nuclear transfer is not a new technique. It was first used in
1952 to study early development in frogs and in the 1980s the
technique was used to clone cattle and sheep using cells taken
directly from early embryos. In 1995, Ian Wilmut, Keith Campbell (see "Cloning Dolly", page 55) and colleagues created live
lambs – Megan and Morag – from fibroblasts (differentiated
cells that have been taken from a foetus) that had been
cultured in the laboratory for several weeks. This was the first
time live animals had been derived from cultured cells and
their success opened up the possibility of introducing much
more precise genetic modifications into farm animals. It was
also the first time that differentiated somatic (that is, non-
embryonic) cells had been sucessfully "reprogrammed" in an
unfertilised egg cell.

In 1996, the Roslin Institute, and collaborators PPL Therapeutics, created Dolly, the first animal cloned from a cell taken
from an adult animal. The announcement of her birth in
February 1997 started the current fascination in all things
cloned. Until then, almost all biologists thought that the cells
in our bodies were fixed in their roles: the creation of Dolly
from a mammary gland cell of a six-year-old sheep showed this
was not the case and the achievement was voted Science
Breakthrough of the Year at the end of 1997.

Progress AD (After Dolly)

At first Dolly was a "clone alone" but foetal fibroblasts were
soon used to produce cloned calves by nuclear transfer, and in

August 1998, a group in Hawaii published a report on the cloning of over fifty mice by nuclear transfer from adult animals. Since then, research groups around the world have reported the cloning of cattle, sheep, mice, goats and pigs. Equally competent groups have had no success in cloning rats, monkeys or dogs.

There are differences in early development between species that might influence success rate. In sheep and humans, the embryo divides to between the 8- and 16-cell stage before nuclear genes take control of development, but in mice this transition occurs at the 2-cell stage. In 1998, a Korean group claimed that they had cloned a human embryo by nuclear transfer but their experiment was terminated at the 4-cell stage and so they had no evidence of successful reprogramming.

Success rates remain low in all species, with published data showing that on average only about 1% of "reconstructed embryos" lead to live births. With unsuccessful attempts at cloning unlikely to be published, the actual success rate will be substantially lower. Many cloned offspring die late in pregnancy or soon after birth, often through respiratory or cardiovascular dysfunction. Abnormal development of the placenta is common and this is probably the major cause of foetal loss earlier in pregnancy. Many of the cloned cattle and sheep that are born are much larger than normal and apparently normal clones may have some unrecognised abnormalities.

The high incidence of abnormalities is not surprising. Normal development of an embryo is dependent on the methylation state* of the DNA contributed by the sperm and egg and on the appropriate reconfiguration of the chromatin structure* after fertilisation. Somatic cells have very different chromatin structure to sperm and "reprogramming" of the transferred nuclei must occur within a few hours of activation of reconstructed embryos. Incomplete or inappropriate reprogramming will lead to dysregulation of gene expression and failure of the embryo or foetus to develop normally or to non-fatal developmental abnormalities in those that survive.

Methylation state: whether or not DNA is methylated. Methylation is the addition of a methyl group to the cytosine at CG sequences and is associated with genes that are not being transcribed..

Chromatin structure: the structure of DNA and histone proteins that make up the basic material of chromosomes.

177

Expression pattern: the postion and timing of gene activation within an organism, with the consequent production of the protein/RNA that they encode.

Improving success rates is not going to be easy. At present, the only way to assess the "quality" of embryos is to look at them under the microscope and it is clear that the large majority of embryos that are classified as "normal" do not develop properly after they have been implanted. A substantial effort is now being made to identify systematic ways of improving reprogramming. One focus is on known mechanisms involved in early development, and in particular on the "imprinting" of genes. Another is to use technological advances in genomics to screen the expression patterns* of tens of thousands of genes to identify differences between the development of "reconstructed embryos" and those produced by *in vivo* or *in vitro* fertilisation.

Limitations of nuclear transfer

It is important to recognise the limitations of nuclear transfer. Plans to clone extinct species have attracted a lot of publicity. One Australian project aims to resurrect the "Tasmanian tiger" by cloning from a specimen that has been preserved in a bottle of alcohol for 153 years and another research group announced plans to clone a mammoth from 20 000 year-old tissue found in the Siberian permafrost. However, the DNA in such samples is hopelessly fragmented and there is no chance of reconstructing a complete genome. In any case, nuclear transfer requires an intact nucleus, with functioning chromosomes. DNA on its own is not enough: many forget that Jurassic Park was a work of fiction.

Other obvious requirements for cloning are an appropriate supply of oocytes and surrogate mothers to carry the cloned embryos to term. Cloning of endangered breeds will be possible by using eggs and surrogates from more common breeds of the same species. It may be possible to clone using a closely related species but the chance of successfully carrying a pregnancy to term would be increasingly unlikely if eggs and surrogate mothers are from more distantly related species. Proposals to "save" the Panda by cloning, for example, would seem to have little or no chance of success because it has no close relatives to supply eggs or carry the cloned embryos.

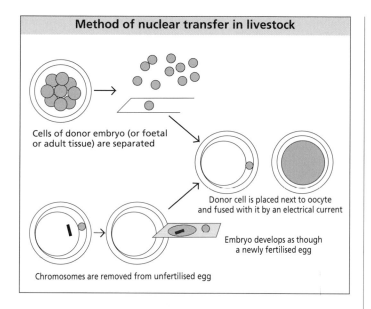

Method of nuclear transfer in livestock

Cells of donor embryo (or foetal or adult tissue) are separated

Donor cell is placed next to oocyte and fused with it by an electrical current

Embryo develops as though a newly fertilised egg

Chromosomes are removed from unfertilised egg

Applications

Nuclear transfer can be viewed in two ways: as a means to create identical copies of animals or as a means of converting cells in culture to live animals. The former has applications in livestock production, the latter provides for the first time an ability to introduce precise genetic modifications into farm animal species.

Cloning in farm animal production

Nuclear transfer can in principle be used to create an infinite number of clones of the very best farm animals. In practice, cloning would be limited to cattle and pigs because it is only in these species that the benefits might justify the costs. Cloned élite cows have already been sold at auction for over $40 000 each in the US but these prices reflect their novelty value rather than their economic worth. To be effective, cloning would have to be integrated systematically into breeding programmes and care would be needed to preserve genetic diversity. It also remains to be shown that clones do consistently deliver the expected commercial performance and are

Deletion:
just as human genes can be spliced into (added to) an animal genome, so unwanted genes could be deleted from it. A transferred nucleus containing this altered genome would then result in an organism where the unwanted genes have effectively been deleted.

healthy and that the technology can be applied without compromising animal welfare.

Production of human therapeutic proteins

Human proteins are in great demand for the treatment of a variety of diseases. Whereas some can be purified from blood this is expensive and runs the risk of contamination by Aids or hepatitis C. Proteins can be produced in human cell culture but costs are very high and output small. Much larger quantities can be produced in bacteria or yeast but the proteins produced can be difficult to purify and they lack the appropriate post-translational modifications that are needed for efficacy *in vivo*.

By contrast, human proteins that have appropriate post-translational modifications can be produced in the milk of transgenic sheep, goats and cattle. Output can be as high as 40 g per litre of milk and costs are relatively low. Nuclear transfer allows human genes to be inserted at specific points in the genome, improving the reliability of their expression and allows genes to be deleted*.

Xenotransplantation

The chronic shortage of organs means that only a fraction of patients who could benefit actually receive transplants. Genetically modified pigs are being developed as an alternative source of organs by a number of companies, though so far the modifications have been limited to adding genes. Nuclear transfer will allow genes to be deleted from pigs and much attention is being directed to eliminating the alpha-galactosyl transferase gene. This codes for an enzyme that creates carbohydrate groups which are attached to pig tissues and which would be largely responsible for the immediate rejection of an organ from a normal pig by a human patient.

Cell-based therapies

Cell transplants are being developed for a wide variety of common diseases, including Parkinson's disease, heart attack, stroke and diabetes. Transplanted cells are as likely to be rejected as organs but this problem could be avoided if the type of cells needed could be derived from the patients themselves. The cloning of adult animals from a variety of cell types shows that the egg and early embryo have the capability of "reprogramming" even fully differentiated cells. Understanding more about the mechanisms involved may allow us to find alternative approaches to reprogramming a patient's own cells without creating (and destroying) human embryos.

Appendix II – Websites

Scientific explanations

Bioethics.net, cloning and genetic
http://www.ajobonline.com/cloning.php

National institute of health: Stem Cells: A Primer
http://www.nih.gov/news/stemcell/primer.htm

Human genome project information cloning fact sheet
http://www.ornl.gov/hgmis/elsi/cloning.html

The cloning of Dolly: a readable explanation from *Science Explained* (January 1998)
www.synapse.ndo.co.uk/science/clone/index.html

Companies and research institutes

PPL Therapeutic:
http://www.bio.org/memberprofile/072435.html

Roslin Institute, Edinburgh:
http://www.ri.bbsrc.ac.uk/library/research/cloning/
http://www.roslin.ac.uk/about/

Wellcome Trust Centre for Human Genetics:
http://www.well.ox.ac.uk/index.html
http://www.wellcome.ac.uk/en/1/awtcon.html

Wellcome/CR UK Institute of Cancer and Developmental Biology
http://www.welc.cam.ac.uk/

International organisations and ethics committees

Council of Europe
http://www.coe.int

Bioethics Division
http://dev.legal.coe.int/bioethics/index_gb.html

International Bioethics Committee (Unesco)
http://www.unesco.org/ibc/index.html

European Union Advisory Group on Ethics
http://europa.eu.int/comm/european_group_ethics/index_en.htm

OECD Bioethics
http://www1.oecd.org/ehs/icgb/

Cloning issues in reproduction, science and medicine
http://www.doh.gov.uk/cloning.htm

French Ethics Committee
http://www.ccne-ethique.org/

Nuffield Council on Bioethics (UK)
www.nuffieldfoundation.org/bioethics/

Euroethics
http://www.gwdg.de/~uelsner/euroeth.htm

Belgian Ethics Committee
http://www.health.fgov.be/bioeth/

Danish Council of Ethics
http://www.etiskraad.dk/english/index.html

German Council of Ethics
http://www.nationalerethikrat.de/

National Ethics Council – Italy
http://www.palazzochigi.it/bioetica/home_eng.html

National Committees for Research Ethics in Norway
http://www.etikkom.no/E/index.htm

Nordic Committee on Bioethics
http://www.ncbio.org/Html/eng_index.htm

Portuguese Ethics Committee
http://www.cnecv.gov.pt/

The President's Council on Bioethics (US)
http://www.bioethics.gov/

Cults

Raelian movement
http://www.clonaid.com/

Articles

"Human cloning" (16.11.2001) A high court ruling means it could be legal for UK scientists to try and clone humans. Derek Brown and Jane Perrone explain the issue
Actuality UK
http://www.guardian.co.uk/Archive/Article/0,4273,4300778,00.html

"A dangerous business"(07.08.2001) Arlene Judith Klotzko
http://www.guardian.co.uk/Archive/Article/0,4273,4234718,00.html

BBC News (2001). "UK enters the clone age."
http://news.bbc.co.uk/hi/english/uk_politics/newsid_1132000/1132034.stm

"Cloning pets"
http://www.telegraph.co.uk/news/main.jhtml?xml=/news/2002/02/15/wcat15.xml

"Cloning: disaster or necessity?" (22.01.2001)
http://www.guardian.co.uk/genes/article/0,2763,426474,00.html

"Human cloning" (22.01.2001)
http://www.guardian.co.uk/theissues/article/0,6512,354116,00.html

"Cloning doctor wants to work in Britain" (28.01.2001)
http://news.telegraph.co.uk/news/main.jhtml?xml=/news/2001/11/05/nbul05.xml

"Yesterday in Parliament" (30.11.2001)
http://news.telegraph.co.uk/news/main.jhtml?xml=%2Fnews%2F2001%2F11%2F30%2Fnpar30.xml

"Experts support human cloning" (16.08.2000)
http://news.bbc.co.uk//hi/english/sci/tech/newsid_881000/881940.stm

"Slouching towards creation"
http://www.pathfinder.com/TIME/cloning/

"Should We Fear Dolly?" (22.02.1997)
http://www.washingtonpost.com/wp-srv/national/longterm/
science/cloning/cloning1.htm

"Human cloning is now 'inevitable' ". (30.08.2000)
http://www.independent.co.uk/story.jsp?story=11480

"Twenty-one arguments against human cloning, and their responses." (01.08.1998)
http://www.geneletter.com/archives/twentyonearguments.html

National Institute for Medical Research
http://www.nimr.mrc.ac.uk/MillHillEssays/1997/cloning.htm

New Scientist special reports:
www.newscientist.com/nsplus/insight/clone/clone.html
http://www.newscientist.com/hottopics/cloning/

"Rabbits join the cloning club" (29.3.2002)
http://news.bbc.co.uk/hi/english/sci/tech/newsid_1899000/
1899477.stm

"Woman pregnant with clone" (05.04.2002)
http://news.bbc.co.uk/hi/english/sci/tech/newsid_1913000/
1913718.stm

Some more links

Karolinska Institute Library Stockholm
http://www.mic.ki.se/Diseases/k1.316.html

Appendix III

Council of Europe Additional Protocol to the Convention for the Protection of Human Rights and Dignity of the Human Being with regard to the Application of Biology and Medicine, on the Prohibition of Cloning Human Beings

Paris, 12 January 1998

The member States of the Council of Europe, the other States and the European Community Signatories to this Additional Protocol to the Convention for the Protection of Human Rights and Dignity of the Human Being with regard to the Application of Biology and Medicine;

Noting scientific developments in the field of mammal cloning, particularly through embryo splitting and nuclear transfer;

Mindful of the progress that some cloning techniques themselves may bring to scientific knowledge and its medical application;

Considering that the cloning of human beings may become a technical possibility;

Having noted that embryo splitting may occur naturally and sometimes result in the birth of genetically identical twins;

Considering however that the instrumentalisation of human beings through the deliberate creation of genetically identical human beings is contrary to human dignity and thus constitutes a misuse of biology and medicine;

Considering also the serious difficulties of a medical, psychological and social nature that such a deliberate biomedical practice might imply for all the individuals involved;

Considering the purpose of the Convention on Human Rights and Biomedicine, in particular the principle mentioned in Article 1 aiming to protect the dignity and identity of all human beings,

Have agreed as follows:

Article 1

1. Any intervention seeking to create a human being genetically identical to another human being, whether living or dead, is prohibited.

2. For the purpose of this article, the term human being "genetically identical" to another human being means a human being sharing with another the same nuclear gene set.

Article 2

No derogation from the provisions of this Protocol shall be made under Article 26, paragraph 1, of the Convention.

Article 3

As between the Parties, the provisions of Articles 1 and 2 of this Protocol shall be regarded as additional articles to the Convention and all the provisions of the Convention shall apply accordingly.

Article 4

This Protocol shall be open for signature by Signatories to the Convention. It is subject to ratification, acceptance or approval. A Signatory may not ratify, accept or approve this Protocol unless it has previously or simultaneously ratified, accepted or approved the Convention. Instruments of ratification, acceptance or approval shall be deposited with the Secretary General of the Council of Europe.

Article 5

1. This Protocol shall enter into force on the first day of the month following the expiration of a period of three months after the date on which five States, including at least four member States of the Council of Europe, have expressed their consent to be bound by the Protocol in accordance with the provisions of Article 4.

2. In respect of any Signatory which subsequently expresses its consent to be bound by it, the Protocol shall enter into force on the first day of the month following the expiration of a period of three months after the date of the deposit of the instrument of ratification, acceptance or approval.

Article 6

1. After the entry into force of this Protocol, any State which has acceded to the Convention may also accede to this Protocol.

2. Accession shall be effected by the deposit with the Secretary General of the Council of Europe of an instrument of accession which shall take effect on the first day of the month following the expiration of a period of three months after the date of its deposit.

Article 7

1. Any Party may at any time denounce this Protocol by means of a notification addressed to the Secretary General of the Council of Europe.

2. Such denunciation shall become effective on the first day of the month following the expiration of a period of three months after the date of receipt of such notification by the Secretary General.

Article 8

The Secretary General of the Council of Europe shall notify the member States of the Council of Europe, the European Community, any Signatory, any Party and any other State which has been invited to accede to the Convention of:

a. any signature;

b. the deposit of any instrument of ratification, acceptance, approval or accession;

c. any date of entry into force of this Protocol in accordance with Articles 5 and 6;

d. any other act, notification or communication relating to this Protocol.

In witness whereof the undersigned, being duly authorised thereto, have signed this Protocol.

Done at Paris, this twelfth day of January 1998, in English and in French, both texts being equally authentic, in a single copy which shall be deposited in the archives of the Council of Europe. The Secretary General of the Council of Europe shall transmit certified copies to each member State of the Council of Europe, to the non-member States which have participated in the elaboration of this Protocol, to any State invited to accede to the Convention and to the European Community.

Sales agents for publications of the Council of Europe
Agents de vente des publications du Conseil de l'Europe

AUSTRALIA/AUSTRALIE
Hunter Publications, 58A, Gipps Street
AUS-3066 COLLINGWOOD, Victoria
Tel.: (61) 3 9417 5361
Fax: (61) 3 9419 7154
E-mail: Sales@hunter-pubs.com.au
http://www.hunter-pubs.com.au

BELGIUM/BELGIQUE
La Librairie européenne SA
50, avenue A. Jonnart
B-1200 BRUXELLES 20
Tel.: (32) 2 734 0281
Fax: (32) 2 735 0860
E-mail: info@libeurop.be
http://www.libeurop.be

Jean de Lannoy
202, avenue du Roi
B-1190 BRUXELLES
Tel.: (32) 2 538 4308
Fax: (32) 2 538 0841
E-mail: jean.de.lannoy@euronet.be
http://www.jean-de-lannoy.be

CANADA
Renouf Publishing Company Limited
5369 Chemin Canotek Road
CDN-OTTAWA, Ontario, K1J 9J3
Tel.: (1) 613 745 2665
Fax: (1) 613 745 7660
E-mail: order.dept@renoufbooks.com
http://www.renoufbooks.com

**CZECH REPUBLIC/
RÉPUBLIQUE TCHÈQUE**
Suweco Cz Dovoz Tisku Praha
Ceskomoravska 21
CZ-18021 PRAHA 9
Tel.: (420) 2 660 35 364
Fax: (420) 2 683 30 42
E-mail: import@suweco.cz

DENMARK/DANEMARK
GAD Direct
Fiolstaede 31-33
DK-1171 COPENHAGEN K
Tel.: (45) 33 13 72 33
Fax: (45) 33 12 54 94
E-mail: info@gaddirect.dk

FINLAND/FINLANDE
Akateeminen Kirjakauppa
Keskuskatu 1, PO Box 218
FIN-00381 HELSINKI
Tel.: (358) 9 121 41
Fax: (358) 9 121 4450
E-mail: akatilaus@stockmann.fi
http://www.akatilaus.akateeminen.com

FRANCE
La Documentation française
(Diffusion/Vente France entière)
124, rue H. Barbusse
F-93308 AUBERVILLIERS Cedex
Tel.: (33) 01 40 15 70 00
Fax: (33) 01 40 15 68 00
E-mail: commandes.vel@ladocfran-
caise.gouv.fr
http://www.ladocfrancaise.gouv.fr

Librairie Kléber (Vente Strasbourg)
Palais de l'Europe
F-67075 STRASBOURG Cedex
Fax: (33) 03 88 52 91 21
E-mail: librairie.kleber@coe.int

**GERMANY/ALLEMAGNE
AUSTRIA/AUTRICHE**
UNO Verlag
Am Hofgarten 10
D-53113 BONN
Tel.: (49) 2 28 94 90 20
Fax: (49) 2 28 94 90 222
E-mail: bestellung@uno-verlag.de
http://www.uno-verlag.de

GREECE/GRÈCE
Librairie Kauffmann
28, rue Stadiou
GR-ATHINAI 10564
Tel.: (30) 1 32 22 160
Fax: (30) 1 32 30 320
E-mail: ord@otenet.gr

HUNGARY/HONGRIE
Euro Info Service
Hungexpo Europa Kozpont ter 1
H-1101 BUDAPEST
Tel.: (361) 264 8270
Fax: (361) 264 8271
E-mail: euroinfo@euroinfo.hu
http://www.euroinfo.hu

ITALY/ITALIE
Libreria Commissionaria Sansoni
Via Duca di Calabria 1/1, CP 552
I-50125 FIRENZE
Tel.: (39) 556 4831
Fax: (39) 556 41257
E-mail: licosa@licosa.com
http://www.licosa.com

NETHERLANDS/PAYS-BAS
De Lindeboom Internationale Publikaties
PO Box 202, MA de Ruyterstraat 20 A
NL-7480 AE HAAKSBERGEN
Tel.: (31) 53 574 0004
Fax: (31) 53 572 9296
E-mail: lindeboo@worldonline.nl
http://home-1-worldonline.nl/~lindeboo/

NORWAY/NORVÈGE
Akademika, A/S Universitetsbokhandel
PO Box 84, Blindern
N-0314 OSLO
Tel.: (47) 22 85 30 30
Fax: (47) 23 12 24 20

POLAND/POLOGNE
Głowna Księgarnia Naukowa
im. B. Prusa
Krakowskie Przedmiescie 7
PL-00-068 WARSZAWA
Tel.: (48) 29 22 66
Fax: (48) 22 26 64 49
E-mail: inter@internews.com.pl
http://www.internews.com.pl

PORTUGAL
Livraria Portugal
Rua do Carmo, 70
P-1200 LISBOA
Tel.: (351) 13 47 49 82
Fax: (351) 13 47 02 64
E-mail: liv.portugal@mail.telepac.pt

SPAIN/ESPAGNE
Mundi-Prensa Libros SA
Castelló 37
E-28001 MADRID
Tel.: (34) 914 36 37 00
Fax: (34) 915 75 39 98
E-mail: libreria@mundiprensa.es
http://www.mundiprensa.com

SWITZERLAND/SUISSE
BERSY
Route de Monteiller
CH-1965 SAVIESE
Tel.: (41) 27 395 53 33
Fax: (41) 27 395 53 34
E-mail: jprausis@netplus.ch

Adeco – Van Diermen
Chemin du Lacuez 41
CH 1807 BLONAY
Tel.: (41) 21 943 26 73
Fax: (41) 21 943 36 05
E-mail: mvandier@worldcom.ch

UNITED KINGDOM/ROYAUME-UNI
TSO (formerly HMSO)
51 Nine Elms Lane
GB-LONDON SW8 5DR
Tel.: (44) 207 873 8372
Fax: (44) 207 873 8200
E-mail: customer.services@theso.co.uk
http://www.the-stationery-office.co.uk
http://www.itsofficial.net

**UNITED STATES and CANADA/
ÉTATS-UNIS et CANADA**
Manhattan Publishing Company
468 Albany Post Road, PO Box 850
CROTON-ON-HUDSON,
NY 10520, USA
Tel.: (1) 914 271 5194
Fax: (1) 914 271 5856
E-mail: Info@manhattanpublishing.com
http://www.manhattanpublishing.com

Council of Europe Publishing/Editions du Conseil de l'Europe
F-67075 Strasbourg Cedex
Tel.: (33) 03 88 41 25 81 – Fax: (33) 03 88 41 39 10 – E-mail: publishing@coe.int – Website: http://book.coe.int